Electronic Transport Theories

From Weakly to Strongly Correlated Materials

Electronic Transport Theories

From Weakly to Strongly Correlated Materials

Navinder Singh

Physical Research Laboratory
Ahmedabad, Gujarat, India

CRC Press
Taylor & Francis Group
Boca Raton London New York

CRC Press is an imprint of the
Taylor & Francis Group, an **informa** business

CRC Press
Taylor & Francis Group
6000 Broken Sound Parkway NW, Suite 300
Boca Raton, FL 33487-2742

Printed on acid-free paper
Version Date: 20160714

International Standard Book Number-13: 978-1-4987-4359-4 (Hardback)

Library of Congress Cataloging-in-Publication Data

Names: Singh, Navinder, 1977- author.
Title: Electronic transport theories from weakly to strongly correlated materials / Navinder Singh.
Description: Boca Raton, FL : CRC Press, Taylor & Francis Group, [2016]
Identifiers: LCCN 2016022637| ISBN 9781498743594 (hardback) | ISBN 1498743595 (hardback) | ISBN 9781498743600 (e-book) | ISBN 1498743609 (e-book)
Subjects: LCSH: Electron transport. | Transport theory. | Nonequilibrium statistical mechanics.
Classification: LCC QC176.8.E4 S44 2016 | DDC 530.4/16--dc23
LC record available at https://lccn.loc.gov/2016022637

Visit the Taylor & Francis Web site at
http://www.taylorandfrancis.com

and the CRC Press Web site at
http://www.crcpress.com

MIX
Paper from responsible sources
FSC www.fsc.org FSC® C013056

Printed and bound in Great Britain by
TJ International Ltd, Padstow, Cornwall

To my parents

Contents

Foreword xi

Preface xiii

List of Figures xxi

List of Tables xxv

1 Introduction and objective of the study **1**

1.1 From Drude-Lorentz to Sommerfeld-Bloch and from Bloch-Peierls to Kubo-Holstein: A historical introduction 1

1.2 Optical absorptivity and reflectivity: brief experimental background 16

1.3 The Drude model at finite frequencies for simple metals . . . 21

1.4 The Lorentz model for simple insulators 25

1.5 Brief discussion of the optical properties of real metals and real insulators 29

1.5.1 Real metals 29

1.5.2 Real insulators 32

1.6 Summary points 34

Bibliography 34

2 The traditional Boltzmann equation based approaches **37**

2.1 Semiclassical model of electron dynamics 38

2.2 Chambers' method for Boltzmann kinetic equation (relaxation time approximation) . . . 42
2.2.1 Obtaining Drude's formula for conductivity from the Boltzmann equation . . . 46
2.2.1.1 DC transport . . . 46
2.2.1.2 AC transport . . . 48
2.3 Beyond the relaxation time approximation (full Boltzmann equation) . . . 49
2.3.1 Physical assumptions for RTA . . . 51
2.3.2 Elastic scattering in an isotropic medium . . . 52
2.4 The Bloch-Boltzmann transport theory . . . 54
2.4.1 The Grüneisen formula . . . 66
2.5 Summary points . . . 69
Bibliography . . . 70

3 Some techniques from nonequilibrium statistical mechanics 71
3.1 Nyquist's thermodynamical arguments and Callen-Welton's fluctuation-dissipation theorem . . . 73
3.2 Kubo's formalism . . . 82
3.2.0.1 Kubo's formula for conductivity . . . 85
3.2.1 Derivation of Drude's formula of conductivity from Kubo's formula . . . 87
3.2.2 Fluctuation-Dissipation Theorem (FDT) from Kubo's formalism . . . 88
3.3 The Drude formula from the Einstein relation . . . 94
3.4 The Drude formula from the Langevin equation . . . 98
3.5 Problem of the Langevin equation . . . 101
3.6 The generalized Langevin equation and the memory function (time dependent friction coefficient) . . . 102
3.7 When can one use the exponential decay of the velocity-velocity correlation function? . . . 105

3.8 Summary points 108
Bibliography 109

4 The Zwanzig-Mori-Götze-Wölfle memory function formalism 111

4.1 The Zwanzig-Mori memory function (MF) formalism 112
4.2 The Götze-Wölfle (GW) formalism 116
4.2.1 The Götze-Wölfle (GW) approximation for the memory function 118
4.2.2 Impurity scattering 120
4.2.3 Phonon scattering 125
4.2.3.1 DC case 129
4.2.3.2 AC case 130
4.3 Summary points 132
Bibliography 133

5 The Kohn-Luttinger theory: Quantum mechanical basis of the Bloch-Boltzmann equation 135

5.1 The assumptions in the traditional kinetic equations 136
5.2 The problems with the assumptions 137
5.3 The model and assumptions 138
5.4 The formalism 139
5.4.1 Ensemble average over the impurities 146
5.5 Why and how are the traditional kinetic theories justified? . 149
5.6 Summary points 150
Bibliography 150

6 Strange metals: A survey 151

6.1 General introduction to the problem 151
6.2 A survey of the experimental situation 153
6.2.1 What are strange metals? 153

6.2.2 Anomalous behavior of DC resistivity 155
6.2.2.1 ab-plane transport 155
6.2.2.2 Optimally doped cuprates 155
6.2.2.3 Underdoped cuprates 160
6.2.2.4 Overdoped cuprates 163
6.2.2.5 c-axis transport 163
6.2.3 Anomalous behavior of AC conductivity 164
6.2.3.1 ab-plane transport 164
6.2.3.2 Optimally doped cuprates 164
6.2.3.3 Underdoped cuprates 167
6.2.3.4 Overdoped cuprates 170
6.2.3.5 c-axis transport 170
6.3 A survey of the theoretical situation 171
6.3.1 The marginal Fermi liquid theory 174
6.3.2 The hidden Fermi liquid theory 178
6.3.3 A brief discussion of other approaches 183
Bibliography 186

7 Electronic transport theories from simple to strange metals: A summary 189
Bibliography 198

8 Supporting material and practice exercises 201
8.1 Appendix A: Proving $f^*_{\mu\nu}(\omega) = f_{\nu\mu}(\omega)$ 201
8.2 Appendix B: Computing $\langle(|H'_{kk'}|^2)^2\rangle_{imp}$ 202
8.3 Appendix C: The concept of quasiparticles 203
8.4 Practice exercises 206

Index 211

Foreword

Transport theory is one of the hardest subjects for theorists. Yet it is a most important one since electrical resistivity, thermal conductivity, and the like are routinely measured in the laboratory. A fruitful dialog between theory and experiment very often must be based on the understanding of transport theory. This book by Professor Singh provides students with a scholarly introduction to the subject. Historical accounts, intuitive ideas, assumptions, justification of the initial assumptions that came later in time, and simple along with sophisticated formalism give an in-depth perspective on the subject that is rarely found in textbooks. The final chapters open up to current research topics that show that transport is a subject where new discoveries are waiting to be made.

André-Marie Tremblay*

*Full Professor and Research Chair in the theory of quantum materials, Physics Department, University of Sherbrooke, Québec, Canada.

Preface

People working in the field of strongly correlated systems (such as high-T_c cuprate superconductors, iron-based superconductors, various heavy electron materials, etc.) know that the electronic transport in these materials is anomalous (e.g., T-linear resistivity over a wide temperature range in the normal states of the mentioned materials in specific parts of the phase diagrams). There is no dearth of refined experiments. But on the theoretical front, the community is facing a fundamental crisis. Traditional theories of the electronic transport based on the Boltzmann equation or on Kubo's linear response theory are not adequate in addressing the anomalous transport. The community is divided on the theoretical front. Some believe that the standard Landau's Fermi Liquid Theory (LFLT) cannot be applied and traditional transport theories have no justification for application to these materials, and on the other hand, others believe that standard LFLT can be applied but we do not have the formalism to treat electronic transport theories due to the presence of strong interactions.

This state of affairs motivated the author of the present treatment to systematically investigate the very foundations of electronic transport theories starting from the Boltzmann kinetic theory to Kubo's landmark theory of the linear response, and to more recent theories. The Boltzmann equation was invented to study molecular dynamics in classical gas theory and its application to electrons in metals is not straightforward. We discuss the subtle points involved when we discuss the semi-classical model. The purpose is to have a thorough presentation of the basic theories (with all the mathematical details for a pedagogical presentation) starting from the traditional theories to more recent theories. The author's expertise in the nonequilibrium statistical mechanics helped to have a transparent presentation of the foundations of the theories of nonequilibrium statistical mechanics as applied to condensed matter systems. The original draft of the book is prepared from lecture notes of a course that was delivered to graduate students at Physical Research Laboratory, Ahmedabad by the author.

The purpose of the book is to provide a thorough background to a person interested in the electronic transport theories in strongly correlated materials that show anomalous transport behavior. The background constitutes the theories of the weakly correlated systems like simple metals. Of course, this state of affairs provides a longer and tedious route, but there is no short-cut!

The organization of the book and a summary of the chapters is as follows. It consists of eight chapters in total. In Chapter 1, a brief historical introduction to the subject of electrical conduction in metals is presented, the classical electromagnetism in the context of reflectivity/absorptivity of a material is reviewed. It is shown how microscopic transport quantities like dielectric function and RF/IF/optical conductivity can be deduced from reflectivity or absorptivity of a material. The case of "simple" metals using Drude model is explicitly worked out. Similarly, a "toy model" introduction to optical/IR properties of insulators is also given using the Lorentz model. A brief introduction to the physical properties of a realistic material is also given. Low frequency and low temperature Holstein mechanism of phonon creation is also discussed. This is to give a flavor of the complexities involved in a realistic system.

Chapter 2, entitled The traditional Boltzmann equation approach, gives a thorough treatment of the theories based on the Boltzmann transport equation. This deals with the electronic transport theories for the weakly correlated systems. In this traditional Boltzmann equation approach, electrons are treated as free and their interaction with impurities or with phonons is considered perturbatively. For metals, this approach holds good due to the application of the Fermi liquid theory electrons in metals can be considered as weakly interacting quasiparticles. The cases of phonon and impurity scattering are worked out in detail. A considerable effort is made to justify (1) the application of the Boltzmann equation to electrons in metals and (2) the Relaxation Time Approximation (RTA). The connection between RTA and quantum statistical effects is made clear. At the end of the chapter a detailed treatment of the Bloch-Boltzmann theory is presented and the famous Grüneisen formula is derived.

Chapter 3 gives techniques from nonequilibrium statistical mechanics are presented. The most important one is Kubo's theory of the linear response. Again, for a "complete" treatment, we start with Nyquist's thermodynamical arguments and Callen-Welton's rigorous formulation of Nyquist's ideas in the form of the Fluctuation-Dissipation Theorem (FDT). A rigorous theory of the transport coefficients was formulated by Kubo along the lines of Nyquist and Callen-Welton. This theory goes beyond the traditional theory of transport (the Boltzmann transport theory). Kubo's theory can be applied to interacting electrons too (no quasiparticle approximation is required for its applicability and it can be applied to non-Fermi liquid materials too, although the mathematical difficulties are enormous). The Kubo theory is worked out in detail for the special case of metals. We have used the original papers of Kubo to work that out. This not only gives a detailed account of the technical issues involved but also, it makes clear the regime of validity and the physical assumptions made in these theories. Almost all the mathematical details are given so that graduate students with "regular" mathematical training can work through it. Other topics from nonequilibrium statistical mechanics are also presented. The Drude formula is also derived from the Einstein relation and the Langevin equation. The problem of the Langevin equation is presented that motivates the introduction of the Generalized Langevin Equation (GLE) and the Generalized Drude Formula (GDF). The GLE involves a time dependent friction coefficient and its Fourier-Laplace transform leads to what is called the Memory Function (MF). The condition under which the MF reduces to a constant is also worked out. This sheds light on when and why the GDF reduces to the simple Drude formula.

In Chapter 4, the Zwanzig-Mori-Götze-Wölfle memory function formalism is presented that takes a different route than the GLE and we show that GDF can be directly obtained from a representation of ac conductivity by Kubo using Zwanzig-Mori projection operators. Next we introduce a very useful workable method for realistic applications–the Götze-Wölfle (GW) memory function formalism. The cases of impurity scattering and phonon scattering in metals are worked out in detail and it is shown how the Holstein mechanism is correctly captured by this method. Also the results previously obtained using the Bloch-Boltzmann equation are re-derived (i.e., the Grüneisen formula can be derived using the memory function formalism).

In Chapter 2 we observed that in formulating the Boltzmann equation for electrons in metals many assumptions and approximations were made. We considered the semi-classical approximation and the Bloch assumption, etc. There is another very subtle assumption that is required, and that we did not mention in the second chapter. This is called the assumption of "random phase," that is, the neglect of the off-diagonal matrix elements of the density matrix (diagonal elements correspond to the occupation probabilities of various states). Chapter 5 is devoted to the analysis of this assumption. There is an elegant and mathematically rigorous formalism due to Kohn and Luttinger which shows, in a simple quantum mechanical setting, that the neglect of the off-diagonal elements of the density matrix in the traditional kinetic equations can be justified. It is the ensemble average over a random distribution of impurities that preferentially suppresses the off-diagonal elements of the density matrix, thus providing a justification of the random phase assumption.

Chapter 6 is devoted to the problems of transport in strongly correlated systems. We consider the case of strange metal observed in the "normal" states of cuprate superconductors. First, the experimental status of the field pertaining to electronic transport is reviewed and then the most popular theories are presented. After listing the prominent ideas, an attempt is made to organize them in terms of microscopic versus phenomenological, and Fermi liquids versus non-Fermi liquids. On the theory side the following theories are reviewed in some detail: the Marginal Fermi Liquid (MFL) theory of Varma et al., and Anderson's Hidden Fermi Liquid (HFL) theory, as these theories enjoy comparatively greater success in dealing with anomalous transport in the "normal" states of high temperature cuprate superconductors. The essential issue of the role of electron-electron interaction in the electronic transport theories is discussed, and at the end other theoretical ideas such as quantum criticality, etc. are discussed in brief.

Since the field is far from being closed, in Chapter 7 we undertake a broader perspective of the whole field starting with a discussion of the rudimentary Drude theory, to a survey of more recent theories. It is like a perspective of the whole book presenting the psychological development of the ideas. The author believes that the book will have permanence (in regard to the first five chapters) but, especially in relation to Chapter 6 on recent theories, one must notice that there is no universally accepted theory for electronic transport in the "normal states" of the above-mentioned materials at present. Thus, Chapter 6 gives a flavor of the current status of electronic transport theories in strange metals. These chapters follow a different mode of presentation as compared to the previous chapters. Instead of detailed calculation, these last chapters provide a perspective of the recent problems.

Finally, the last chapter contains some supporting material like appendices and some practice exercises.

Many friends and colleagues gave their suggestions and comments. The author would like to thank P. W. Anderson for his critical comments on Chapter 6 on strange metals, and some of his suggestions are incorporated. He would like to thank T. V. Ramakrishnan for his suggestion about classifying the various theories presented in Chapter 6. The author's greatest gratitude goes to Bharti Kindra and Bhavesh Chauhan for their scrutiny of the manuscript, and suggesting many corrections and improvements. He would also like to thank Namit Mahajan, D. P. Dewangan, Nabyendu Das, and Natasha Shiggon for carefully reading portions of the manuscript and suggesting improvements/corrections. Auther's sincere thanks are also due to Aastha Sharma (editor, CRC press) for her constant support and reminders to meet the deadlines, and to Michele Dimont (CRC press) for suggesting numerous corrections in the proofs which has improved the text considerably. At the end, he would like to express his regards to his wife, Manpreet Kaur, for her patience, and for the time that the writing of this book took from their lives.

The author's advice to students interested in the topic is that they must work through the mathematical details of the formalism by themselves. All the essential steps are given. This is the only route to a deeper understanding of the electronic transport in strongly correlated materials.

Navinder Singh,

Physical Research Laboratory, Ahmedabad.

The champions of the theories of electrical conduction

(a)

(b)

(c)

(d)

(e)

(f)

(g)

(h)

(i)

(a) **Paul Drude** (July 12, 1863 – July 5, 1906) developed the kinetic theory of metallic conduction in 1900 by treating electrons as classical particles bumping off the immobile ion cores. (b) **Hendrik Lorentz** (July 18, 1853 – Feb. 4, 1928) first time (in 1905) used the classical Boltzmann equation to derive the Drude result. (c) **Arnold Sommerfeld** (Dec. 5, 1868 – April 26, 1951) resolved the problems of the classical Drude model using then newly discovered quantum mechanics. He basically considered the fermionic nature of electrons (in 1928), that is, by replacing the Maxwell-Boltzmann distribution in Lorentz's theory by the Fermi-Dirac distribution. (d) **Felix Bloch** (Oct. 23, 1905 – Sept. 10, 1983) developed quantum description of electrical conduction using what is now called the Bloch-Boltzmann equation including the electron-phonon scattering (in 1930) thus explaining the temperature dependence of electrical resistivity. (e) **Alan Wilson** (July 2, 1906 – Sept. 30, 1995) studied the mechanism of electrical conduction in semiconductors (1932) and elucidated the Bloch theory. (f) **Lev Landau** (Jan. 22, 1908 – April 1, 1968), along with his many seminal contributions in solid state physics and many other areas, developed the notion of quasiparticles and the Fermi liquid theory which provided the conceptual justification of the application of the Boltzmann-Bloch theory to electrical conduction in metals. (g) **Rudolf Peierls** (June 5, 1907 – Sept. 19, 1995) formulated the Bloch-Boltzmann kinetic theory on a rigorous foundation including the non-equilibrium nature of phonon distribution (and the phonon drag effects) and worked out the conditions of the applicability of the Bloch-Boltzmann theory. He also worked out the role of phonon-phonon interaction in the mechanism of heat conduction. (h) **Ryogo Kubo** (Feb. 15, 1920 – March 31, 1995) developed an alternative method to the Bloch-Boltzmann equation in 1957. The Bloch-Boltzmann equation can only be applied to weakly interacting quasiparticles while the Kubo method can be applied to strongly interacting particles. In Kubo's method, the transport coefficients are directly expressed in terms of corresponding correlation functions which are computable (in some situations). Thus there are two main methods for attacking transport problems (1) the Boltzmann equation method and (2) Kubo's linear response theory. (i) **Ted Holstein** (Sept. 18, 1915 – May 8, 1985) developed a rigorous theory of electrical transport involving electron-phonon scattering. Starting from the Kubo formula he developed a rigorous transport equation known as Holstein-Boltzmann equation (in 1964).

List of Figures

1.1 A hypothetical model of electrical conduction in metals by Weber. 3
1.2 The group photo of the 4th Solvay conference on electrical conduction in materials (1924). 9
1.3 A historical sketch of important contributions. 12
1.4 A sketch of an apparatus for measuring reflectivity/absorptivity of a material. 17
1.5 Schematic depiction of the reflectance spectra of Ag and Cu. Vertical dot-dashed lines show the regimes where d-band transitions are induced. 31

2.1 The semiclassical model of the electron dynamics. The lattice constant is the smallest length scale and the wavelength of the external field is the largest one. The size of the electron wavepacket lies in between. 41
2.2 A segment of the trajectory of the phase point in the $\mu-$phase space (phase space of a single electron of 6 dimensions). The phase trajectory is governed by the semiclassical equations of motion. 43
2.3 The phonon emission and absorption processes. 56
2.4 Diagram for the evaluation of q integral. Electric field is along the $x-$direction. 63

3.1 (a) Two resistors, both at the same temperature, are connected with each other. (b) An arrangement to show that in any frequency interval equal power flows from both sides (power flow balance). (c) Standing modes in a shunted transmission line. 74
3.2 Diffusing colloidal particles in a suspension. 95
3.3 The volume above the shaded area due to the surface $\langle u(t_1)u(t_2)\rangle$ is half the volume above the total area t^2. The function $\langle u(t_1)u(t_2)\rangle$ is symmetric around the line $t_1 = t_2$. . 96

5.1 System is composed of a cubical box of volume V that contains randomly distributed impurities, and it also contains a dilute electron gas. To maintain a steady current flow, periodic boundary conditions are imposed on the faces. 139
5.2 Periodic boundary conditions to mimic continuity and thermodynamic limit of a system. 143

6.1 A typical phase diagram of cuprates. 154
6.2 DC resistivity at optimal doping (a sample). Squares are the data points for $La_{2-x}Sr_xCuO_4$ at $x = 0.21$[10], and the dotted line is a linear fit to the data above the superconducting transition. Filled circles are the data points for $Bi-2201$ at optimal doping [11] and the solid line is a linear fit to the data. 156
6.3 T-linear component of resistivity is always present in the regime above the SC dome. The shaded area implies the presence of the $T-$linear component. This points toward an important link between the mechanism of T-linear resistivity and the mechanism of Cooper paring in cuprates. 157
6.4 (a) Resistivity of metals and (b) resistivity of cuprates at optimal doping. 159

6.5 At $T \simeq T^*$ the resistivity shows a change in the slope which roughly coincides with what is observed in NMR experiments. 161
6.6 The scaling relation of resistivity in underdoped cuprates. . 162
6.7 The defining characters of the strange metals regime. 165
6.8 The high frequency part of the conductivity is temperature independent. 167
6.9 (a) At $T < T^*$ a gap-like feature appears in the optical conductivity with a dip at $\omega \sim 500\ cm^{-1}$. (b) A corresponding signature appears in the generalized Drude scattering rate when the temperature is brought below T^*. 168
6.10 The famous MIR band appears as the material is doped away from the insulator state. With increasing doping this feature shifts to lower frequencies and disappears at overdoping. . . 169
6.11 The mystery of the MIR band and its plausible explainations. 169
6.12 The c-axis IR conductivity showing the opening of a pseudogap below $T^* \simeq 300K$. 170
6.13 The "bottleneck" mechanism of the HFLT. 181

8.1 Schematic spectral functions in two systems A and B. . . . 206
8.2 A Mathematica program that shows that the off-diagonal goes as $\sqrt{N}$. 208

List of Tables

1.1 Reflectance and absorption coefficient for simple metals that obey the Drude model 25

1.2 Reflectance and absorption coefficient for simple insulators that obey the Lorentz model. Γ is the Lorentz scattering rate, ω_p is the plasma frequency, and ω_0 is the harmonic frequency. 29

6.1 ab-plane resistivities of various cuprates per CuO_2−plane are of the same order of magnitude. This observation clearly shows that electronic transport is mainly controlled by the CuO_2−planes. 156

Chapter 1

Introduction and objective of the study

"At the moment we have the feeling that we are at a dead end, with the old theories showing themselves more and more powerless to pierce the darkness that surrounds us from all sides.[...] What will be the result of these meetings? I dare not predict it, not knowing what surprises may be in store for us. But, as it is wise not to count on surprises, I admit that it is very likely that we will contribute only little to the immediate progress."

H. A. Lorentz*

1.1 From Drude-Lorentz to Sommerfeld-Bloch and from Bloch-Peierls to Kubo-Holstein: A historical introduction

It is a matter of common experience that when a potential difference is impressed on a metallic sample current starts to flow across it. If the potential difference is not too strong, one is in the linear regime and current induced in the sample is proportional to the applied voltage and the proportionality con-

*From his opening address of the first Solvay conference held at Brussels from Oct. 30 to Nov. 3, 1911. For details refer to: A. J. Kox, *Arch. Hist. Exact Sci.* (2013) 67:149-170.

stant is the resistance of the sample (the famous Ohm's law). The resistance is a measure of the impediment of the electronic motion under an external bias. This impediment in simple metals is either due to impurity or due to phonon scattering, or both mechanisms working together. At very low temperatures the impurity scattering dominates over the phonon scattering and at high temperatures it is vice versa. The notion of low and high temperatures is with respect to a characteristic temperature called the Debye temperature. Resistance of a given sample varies with temperature and this temperature dependence, which can be measured in the laboratory, has hidden information regarding microscopic electronic scattering mechanisms. From a measurement of resistance or resistivity (which is independent from the dimensions of the sample) one can infer about microscopic scattering mechanisms. To accomplish this, one has to develop a theoretical model for microscopic scattering mechanisms.

The first notable theoretical model that connects the resistivity with a microscopic mechanism of electron scattering via localized ion cores was given by Paul Drude in 1900. Even before Drude, Weber in 1875 made an attempt to understand the mechanism of conduction in metals. The electron was not discovered at that time but Michael Faraday's electrochemical experiments showed that charge is transported by some sort of carriers in ionic solutions. Building on this, Weber imagined that metals consist of unstable and mobile molecules which can absorb and emit charge carriers (Figure 1.1). Under the action of an applied potential difference, molecules would emit charge carriers, and then these carriers would "flow" in metal under the action of bias and would get absorbed by the next molecule (Figure 1.1). Thus, electrical conduction would happen through these emission and absorption processes. The theory was a kind of speculation and did not yield any concrete result.

Then, in 1897, J. J. Thomson discovered the electron and struggle to understand the constitution of atoms ensued. Three years after Thomson's discovery, in 1900, Paul Drude came up with a better-founded model for electrical conduction in metals. Taking inspiration from the kinetic theory of gases, Paul Drude treated the then-newly-discovered electrons as classical particles, much like gas molecules, and introduced their scattering with localized ion cores as a mechanism for their impediment. Along with many interesting results, he gave a theoretical description of an empirical fact discovered in 1853 by Wiedemann and Franz, which states that the ratio of electrical and thermal conductivities at a given temperature is the same for all metals. Below we derive the Drude result.

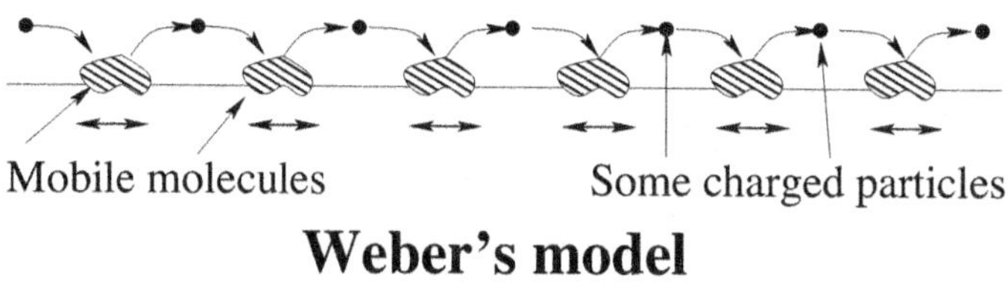

FIGURE 1.1: A hypothetical model of electrical conduction in metals by Weber.

Paul Drude's model is based on the following assumptions:

1. Electrons are noninteracting classical particles (much like gas molecules) and the mechanism of scattering is the electron-ion interaction.
2. Immediately after a collision with an ion core, the electron's velocity is randomized.

When there is no external electric field, the electronic velocities are random and pointing in all directions with no net current flow. With the application of external electric field there will be a systematic velocity component (the drift velocity) in a direction opposite to the applied electric field (due to negative charge on the electron). Let us assume that the electric field (E_x) is applied in the $x-$direction. We want to find the average velocity at an arbitrary time instant t. Let us assume that an electron, say electron number 1, is freely accelerating for time t_1 after its last collision with an immobile ion. According to assumption 2 above, its velocity was random just after its last collision. Another electron, say number 2, is accelerating for time t_2 after its last collision. Similarly the ith electron would have accelerated for time t_i after its last collision. Then the velocity component of an ith electron which accelerates for time t_i is given by

$$u_x^i(t) = u_{0,x}^i - \frac{eE_x}{m} t_i, \tag{1.1}$$

where $u_{0,x}^i$ is the $x-$component of the random velocity just after the collision. Notice that the time t_i varies from electron to electron. An electron which had its last collision just before the time instant t (at which we want to find the average velocity) would have accelerated for a shorter time interval, while an electron that had its last collision quite a while before the time instant t would have accelerated for a comparatively longer time. Let us introduce an average time for acceleration (ATFA) over all the electrons:

$$\tau = \frac{1}{N} \sum_{i=1}^{N} t_i. \tag{1.2}$$

Here N is the total number of electrons in the sample of volume V. Thus, the average velocity from Equation (1.1) is

$$\langle v_x \rangle = \frac{1}{N} \sum_i (u_{0,x}^i - \frac{eE_x}{m} t_i) = -\frac{eE_x}{m} \tau, \tag{1.3}$$

where $\sum_i u_{0,x}^i = 0$ (by assumption 2 above). The average current density takes the form

$$J_x = -en\langle v_x \rangle = \frac{ne^2 \tau}{m} E_x. \tag{1.4}$$

From Ohm's law $J_x = \sigma_{xx} E_x$, the $x-$component of the conductivity is given by

$$\sigma = \frac{ne^2 \tau}{m}. \tag{1.5}$$

which is the famous Drude formula for DC conductivity. An important point that one must recognize here is that the time for acceleration t_i for an ith electron varies drastically from electron to electron (depending upon the instant of its last collision from time t). The average time introduced above (τ) has a meaning only in the ensemble sense, it does not have any meaning for just one electron.

To derive the Wiedemann-Franz law we need to calculate the thermal conductivity also. In the calculation of thermal conductivity, we will need the average time between two successive collisions. There are two different time scales: (1) ATFA, and (2) the average time between two successive collisions. If ATFA is τ, then the average time between two successive collisions is 2τ*.

*This important point is missed in the popular textbook: *Solid State Physics* by Ashcroft and Mermin, but taken into account in an older book, *The Theory of Metals* by A. H. Wilson. The proof is simple: To find the relationship between the ATFA and the Average Time between Two Successive Collisions (ATTSC) let us assume a uniform distribution of electrons over a 1-D lattice of ions with lattice constant a. Let us divide the distance a into m bins with each bin of width a/m. As the electron density is uniformly spread out over the lattice, each bin on average will have an equal number of electrons. Let us assume that each bin has n electrons on average and assume that electrons drift toward the right direction under the action of applied field. The electrons in the bin that is the farthest from the lattice ion on the right side will take an average time 2τ, say, to reach there. Then, the electrons in the next bin will take time $2\tau - \frac{2\tau}{m}$, and from the third bin to the ion on the right will take time $2\tau - 2\frac{2\tau}{m}$ and so on. The electrons from the $m-1$th bin which is near to the ion on the right will take time $2\tau - (m-1)\frac{2\tau}{m}$, and the electrons from the mth bin which is touching the ion on the right will take zero time. Thus, the average time for which the electron undergoes acceleration (ATFA) before colliding with the ion is given by

$$ATFA = \frac{(2\tau) + (2\tau - \frac{2\tau}{m}) + (2\tau - 2\frac{2\tau}{m}) + ...(2\tau - (m-1)\frac{2\tau}{m}) + (2\tau - 2\tau)}{m}, \tag{1.6}$$

which can be simplified as

$$ATFA = 2\tau - \frac{1}{m}(\frac{2\tau}{m} + 2\frac{2\tau}{m} + ... + (m-1)\frac{2\tau}{m}). \tag{1.7}$$

With the sum of AP in parenthesis, the average time for acceleration becomes

$$ATFA = 2\tau - \frac{m-1}{m}\tau, \tag{1.8}$$

which in the large m limit gives $ATFA = \tau$. Thus if ATTSC is 2τ, then ATFA is τ. One can argue that due to acceleration the uniform distribution of times for all the bins is questionable. The fact is that the change in the velocity due to a typical electric field is

For the calculation of thermal conductivity within Drude's formalism, again consider a one-dimensional model. Let us assume that left end of a 1-D metallic conductor is kept at higher temperature with respect to the right end. This induces a temperature gradient along the length of the conductor (the temperature difference is maintained by attaching the conductor with appropriate thermal baths). We assume that the phenomenological Fourier's law of heat conduction is obeyed

$$J_x^q = -\kappa \frac{dT}{dx}, \qquad \text{in 1D}, \quad \mathbf{J}^q = -\kappa \nabla T, \qquad \text{in 3D}. \tag{1.9}$$

where κ is heat conductivity. Let us denote the lattice sites with x_i where x_i is the position of the ith ion core measured from the left end, and the lattice constant is denoted as a. Heat current density that flows through an ion at x_i from left to right (hotter end to colder end) is given by

$$J_x^q = \frac{1}{2} n v_x(x_i - a)\xi(T(x_i - a)) - \frac{1}{2} n v_x(x_i + a)\xi(T(x_i + a)), \tag{1.10}$$

where $\xi(T(x_i))$ is the energy of an electron emanating from an ion at x_i after a collision with it and $v_x(x_i)$ is its velocity component along the $x-$direction at that site. The first term in the above equation depicts the energy flowing toward right side per unit time per unit surface area at position x_i. The factor of half in front of n captures the fact that $n/2$ (out of n) number of electrons per unit volume are moving toward right (no electrical current flow within this approximation). Similarly the second term depicts the heat current density flowing toward the left (thus with the minus sign). One should check dimensions for consistency.

If c_e is the heat capacity per electron, then $\xi(T) = c_e T$. Substituting this in the above equation and as $x_i >> a$ (x_i is the macroscopic length scale measured from left end (say) while a is the microscopic one), one can Taylor

negligible as compared to the magnitude of the velocity itself: $\Delta v_x = \frac{eE_x}{m}\tau \simeq 10\ cm/sec$ (for the typical values $E_x \sim 10^2\ Volts/m,\ \ \tau \sim 10\ femtoseconds$), while $v_x \sim 10^5\ m/sec$ (using classical equipartition theorem $\frac{1}{2}mv_x^2 = \frac{1}{2}k_B T$). This yields $\Delta v_x << v_x$. Thus, the above argument is watertight.

expand the above equation with the leading order term†

$$J_x^q \simeq -v_x c_v a \frac{dT}{dx}. \tag{1.11}$$

Here $c_v = nc_e$ is the heat capacity per unit volume. We showed in the calculation of electrical conductivity that if the average time for acceleration is τ, then the average time between two successive collisions is 2τ. Thus, $a = v_x 2\tau$ (where a is the distance between two adjacent ions). Also generalize the above calculation to 3D in an isotropic system: $v_x^2 = \frac{1}{3}v^2$ (i.e., $v_x = v_y = v_z$). The current density can be written as

$$\mathbf{J}^q = -\frac{2}{3}v^2 c_v \tau \nabla T. \tag{1.12}$$

Comparing with the Fourier law gives the thermal conductivity

$$\kappa = \frac{2}{3}v^2 c_v \tau. \tag{1.13}$$

From Equations (1.5) and (1.13)

$$\frac{\kappa}{\sigma} = \frac{2}{3}\frac{mv^2 c_v}{ne^2} = 3\left(\frac{k_B}{e}\right)^2 T. \tag{1.14}$$

In the above we used the classical equipartition theorem: $\frac{1}{2}mv^2 = \frac{3}{2}k_B T$ and the classical result for the heat capacity of electron gas: $c_v = \frac{3}{2}nk_B$. The above equation is known as the Wiedemann-Franz law and the numerical factor $3\left(\frac{k_B}{e}\right)^2$ is called the Lorentz number. Its value turns out to be $\simeq 2.5 \times 10^{-13}$ $(erg/esu - K)^2$ which is very close to the experimental value $\simeq 2.7 \times 10^{-13}$ $(erg/esu - K)^2$.

This excellent agreement between theory and experiment was an apparent success of the Drude theory. But it was soon realized that there was not just one but several problems with the Drude theory. Something serious was missing. The first offset to the Drude theory was its failure to account for the temperature dependence of electrical and thermal conductivities (Equations

---

†There will be another weaker term $-nac_e T\frac{dv_x}{dx}$ along with the Fourier law which we neglect as the Fourier law is assumed to hold good for the present case [1].

(1.5) and (1.13)). Hendrik Lorentz investigated the problem in 1905 from a more rigorous but still classical point of view using the then newly discovered Boltzmann equation. No significant improvement over the Drude theory was achieved. The Lorentz number from his theory turns out to be $2\left(\frac{k_B}{e}\right)^2$ which is in worse agreement with the experiment. The problem of the temperature dependence of heat and electrical conductivities was not resolved either.

Later on, in 1912, Debye pointed out that the heat capacity of metals can be satisfactorily accounted for by considering only the contribution from the lattice vibrations (phonons). If there are N atoms in a metallic sample of unit volume, then there are $3N$ degrees of freedom, and heat capacity per degree of freedom is $\frac{1}{2}k_B$. Each atom has both kinetic energy and potential energy; thus, the heat capacity per atom is k_B and the heat capacity of the solid per unit volume is $3Nk_B$, which showed good agreement with experiments at high temperatures (greater than Debye temperature). If one uses Drude's ideas and treats electrons as classical particles and applies the above argument to the electron gas, then heat capacity per unit volume for the electron gas turns out to be $\frac{3}{2}N_e k_B$ where N_e is the number of electrons in the unit volume and a factor of half comes due to the absence of potential energy in the free electron gas. Practically $N_e \sim N$ (number of ions are roughly of the order of the number of free electrons), thus electrons should cause a sizable contribution to heat capacity (of the order of lattice heat capacity) and this electronic contribution was never found in the experiments.

In the second decade of the last century, the situation became very confusing as on one hand the classical theory predicted the Wiedemann-Franz law so nicely, while on the other hand it failed terribly to account for temperature dependence of conductivities and the electronic heat capacity. The state of affairs was well recorded in the report of the 4th Solvay Conference on Physics held at Brussels, Belgium from April 24 to 29, 1924. It was organized by H. Lorentz and along with P. W. Bridgman, P. Debye, H. K. Onnes, P. Langevin, E. Schrödinger, Madame Curie, and many other distinguished

FIGURE 1.2: The group photo of the 4th Solvay conference on electrical conduction in materials (1924).

physicists of the time participated (Figure (1.2)). The founding fathers clearly recognized the deeper problems but the solution was not available as the holy grail of the problem, the Fermi-Dirac distribution, was not in place at that time.

In 1925, Pauli discovered the exclusion principle, which permits at most one electron to occupy any single electron level. Building on the exclusion principle, in 1926 Fermi and Dirac independently developed a new form of statistics, which reduced to the Maxwell-Boltzmann form in the low electron density limit. The Fermi-Dirac distribution presented a radically different form of the electron distribution in metals as compared to the Maxwell-Boltzmann distribution.

In the years 1927 and 1928, A. Sommerfeld reinvestigated the Drude-Lorentz theory but by replacing the Maxwell-Boltzmann distribution with the Fermi-Dirac distribution. He got immediate success on two accounts; one related to Lorentz number and the other related to electronic heat capacity.

He assumed the existance of mean acceleration time τ as in the Drude theory formalism (or relaxation time in the Lorentz theory using the Boltzmann equation) and treated it as a parameter in the model without going into its microscopic justification. From the Drude-Lorentz approach he obtained the same expressions for electrical and thermal conductivities, i.e., Equations (1.5) and (1.13). The important point is that, with the use of Fermi-Dirac statistics, heat capacity of the electrons takes the form $c_v = \frac{\pi^2}{2} n k_B (k_B T/E_F)$ instead of the classical value $(3/2) N_e k_B$. The Fermi-Dirac distribution introduces an important energy scale in the problem, i.e., the Fermi energy E_F. And the electron velocity is not given by the classical formula $\sqrt{3k_B T/m}$ but by the correct quantum mechanical expression $\sqrt{2E_F/m}$. This leads to the Lorentz number L given

$$L = \frac{\pi^2}{3} \left(\frac{k_B}{e} \right)^2, \tag{1.15}$$

which stays in an excellent agreement with the experiment. The most important point of the Sommerfeld theory is that it predicts the correct value of the electronic heat capacity which is negligibly small as compared to its classical value thus justifying Debye's observation that the heat capacity of metals at ordinary temperatures is predominately due to phonons.

This was the turning point. The problems that baffled the pioneers like Lorentz and others over two decades were partially resolved at one fell swoop. We notice that Sommerfeld's theory resolved one big obstacle, i.e., the electronic heat capacity issue. But it did not resolve the problems of the temperature dependencies of electrical and thermal conductivities. The solution of this problem was hidden under the complexities of the scattering time τ. This problem was resolved by Bloch in 1928 when he investigated the scattering of electrons by lattice vibrations, and along with that he introduced the very core of the subject of the quantum theory of metals (i.e., the Bloch theorem [1]). We will study the Bloch theory which is now called the Bloch-Boltzmann theory in the next chapter after introducing the Boltzmann equation.

Before we present the summary of this very interesting early history, let us examine why Drude's theory was able to predict the correct value of the Lorentz number. Here it is just a coincidence. In the heat conductivity formula (Equation (1.13)) we encounter the product of heat capacity and square of the electronic velocity, i.e., $c_v v^2$. The classical value of the electronic heat capacity is 100 times larger than the correct quantum mechanical result and the square of the electronic velocity is 100 times smaller than the correct quantum mechanical result. Thus, these factors cancel out in the product giving the correct value of the heat conductivity and thus the ratio with electrical conductivity yielding the correct value of the Lorentz number. We notice from the above discussion that it was a "lucky" circumstance which highlighted its temporal success but as we have seen from the above, the Drude theory completely failed on other accounts.

We close this account of the early history by summarizing it in Figure (1.3). F. Bloch introduced the idea of Bloch waves. He solved the Schrödinger equation in a *perfect periodic lattice* and found that electrons do not scatter; rather, they form an extended wave throughout the lattice called the Bloch wave.

To investigate transport he used the Boltzmann equation. The scattering mechanism that Bloch considered was the electron-phonon scattering. He computed the transition rate of electrons scattering from one Bloch state to another using Dirac's time dependent perturbation theory and used it into the collision term of the Boltzmann equation. The resulting equation is now popular as the Bloch-Boltzmann equation. The solution, in the high temperature limit, predicted the correct temperature dependence of conductivities and also the Wiedemann-Franz law was reproduced.

Later on, in 1933, Grüneisen systematically studied the Bloch-Boltzmann equation both in the low temperature limit $T << T_D$ and in the high temperature limit $T >> T_D$ (where T_D is the Debye temperature) and came up with the formula of electrical resistivity

$$\rho \propto \left(\frac{T}{T_D}\right)^5 J_5(T/T_D), \qquad J_5(x) = \int_0^x dx \frac{x^5}{(e^x - 1)(1 - e^{-x})}. \tag{1.16}$$

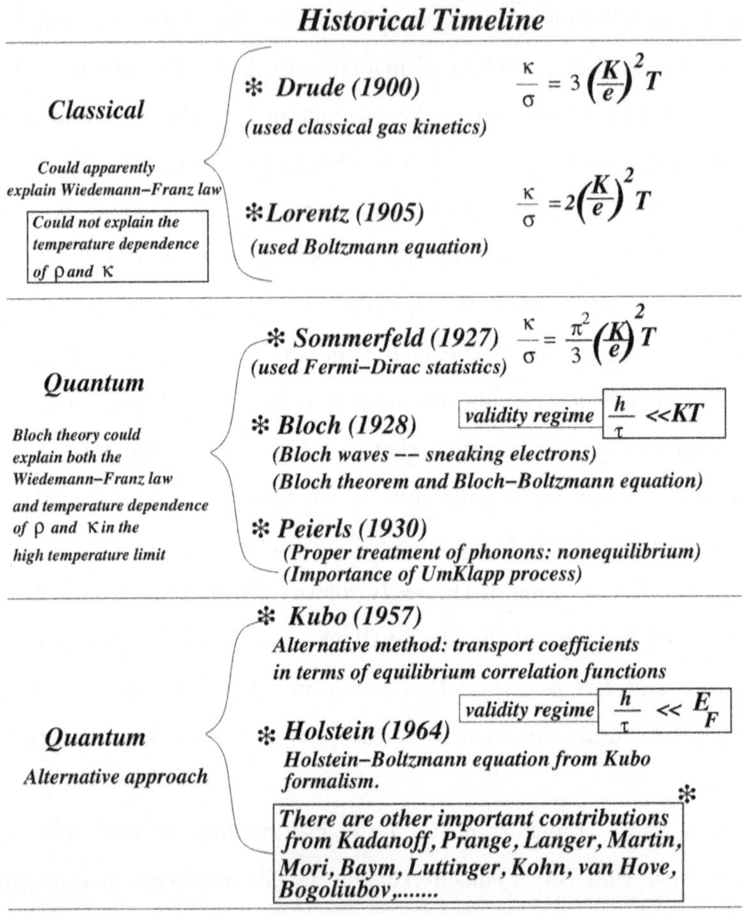

FIGURE 1.3: A historical sketch of important contributions.

which is now called the Grüneisen formula and predicted $\rho \propto T$ in the limit $T >> T_D$ and $\rho \propto T^5$ in the limit $T << T_D$. This stands in a reasonably good agreement with empirical facts in metals.

There were many problems with the Bloch-Boltzmann theory. No theory is a perfect theory! There are two kinds of problems: (1) Internal logical problems and (2) discrepancies with experiments at low temperatures.

The first kind of problems are related to internal logical contradictions due to the semiclassical nature of the theory. If τ is the relaxation time in the Bloch-Boltzmann theory and is roughly the average time between two successive collisions, and if $k_B T$ is the total range of energy that electrons can have (as electrons in the energy zone of width $k_B T$ around the Fermi energy actively participate in conduction), then for the semiclassical argument to hold the energy uncertainty introduced due to finite τ must be much less than $k_B T$:

$$\frac{\hbar}{\tau} << k_B T. \tag{1.17}$$

As $k_B T \simeq 1/40\ eV \simeq 10^{-20}\ Joules$(at room temperature) and $\tau \simeq 10\ femtosec$, then $\frac{\hbar}{\tau} \simeq 10^{-20}\ Joules$. Thus the above relation is not justified. Actually at room temperature, τ nearly saturates the quantum bound. However, the correct bound [10] that applies to the metals is

$$\frac{\hbar}{\tau} << E_F. \tag{1.18}$$

This is true for elastic scattering (but it is generally valid even for the case of electron-phonon scattering. In electron-phonon scattering, the phonon energies are very tiny as compared to electron energies, and thus the scattering is approximately elastic).

The second kind of problems are related to the low temperature behavior of the Bloch-Boltzmann theory. At low temperatures there are discrepancies between theory and experiments. Fermi surface peculiarities start to play an important role (refer for details [3, 4]).

In the 1930s, Peierls noted that (1) to dissipate momentum from electron subsystem to the lattice bath the Umklapp processes play an essential role, and (2) energy is first absorbed by the electrons from the field and then it is dissipated to the phonon bath through electron-phonon interactions. During this process, phonons may not remain in equilibrium and one must consider the nonequilibrium distribution of the phonons explicitly[‡]. He developed

---

[‡]Bloch assumed that phonons very quickly redistribute their energy among themselves and reach an equilibrium distribution over a time scale much shorter than the time scale over which the energy is transferred from the electronic subsystem to the phononic subsystem. This is called the Bloch assumption, and it is discussed in detail in the next chapter.

what is now called the Peierls-Bloch-Botzmann equations, which consider the non-equilibrium nature of the phonons explicitly (refer to [5] and references therein).

The approach starting from Drude-Lorentz and culminating in the Bloch-Boltzmann theory is called the traditional approach. This traditional approach is centered around the application of the Boltzmann equation to electrons in metals. We see from the above history that with the advent of quantum mechanics the classical approach of Drude-Lorentz was replaced with the semi-classical Bloch-Boltzmann theory which better accounted for the observed phenomena. The later developments more or less followed this line of approach. And even up to the 1990s, the theories of hot electron relaxation in metals and metal films were based on the Bloch-Boltzmann theory. Until today, it remains the approach of choice (due to its conceptual simplicity). The main drawback of the Bloch-Boltzmann or the traditional approach is that it is applicable only to weakly interacting particle systems (refer to Chapter 2).

In 1957, Kubo and others introduced an alternative approach to transport and non-equilibrium problems. One starts with the Liouville-von Neumann equation for the total density matrix and treats the field-matter Hamiltonian as a perturbation. Expectation value of an operator under study is found by using the linear response theory. This approach is discussed in Chapter 3. The end result is that the transport coefficients are expressed in terms of the time integrals of corresponding equilibrium correlation functions. The main advantage of Kubo's approach over the traditional approach is that it is fully quantum in principle and it can be applied to interacting systems (the traditional approach requires weakly interacting particles). The only restriction is that it works in weak dis-equilibrium situations whereas the Boltzmann equation can be applied to strong dis-equilibrium situations like hot electrons in metals.

Another major work appeared in 1964. Ted Holstein, starting with the Kubo theory, investigated very rigorously the case of electron-phonon interactions. The problems of the uncontrolled perturbation theories in Bloch-Boltzmann approach were resolved by finding appropriate perturbation parameters. He found that the proper perturbation parameter is not the electron-phonon interaction as used in Bloch-Boltzmann theory but the ratio of sound speed with Fermi velocity. The end result is the Holstein-Boltzmann equation which takes into account the non-equilibrium nature of phonons and the phonon drag effects. From his analysis, it turns out that the correct condition for the applicability of Boltzmann equation is

$$\frac{\hbar}{\tau} << E_F, \tag{1.19}$$

not the condition (1.17). The above condition is very well satisfied by metals.

There are other important contributions from investigators like Kadanoff, Baym, Prange, Langer, Martin, Mori, Luttinger, Kohn, Bogoliubov, van Hove, Mahan, and many others. The approaches of Kadanoff-Baym, Langer, Martin, and Mahan are based on Green's function methods. van Hove investigated the problem from a master equation approach. The literature is very rich. Some of the approaches will be discussed in this book. The interested reader should browse the literature [6].

In the next section we will not follow the historical line and move on to the finite frequency transport, i.e., optical spectroscopic. We are going to discuss reflectivity/absorptivity of metals/insulators in the context of simple Drude and Lorentz models, and then compare that with the optical properties of real metals and real insulators.

1.2 Optical absorptivity and reflectivity: brief experimental background

In this section we study IR/optical conductivity using the simple Drude model. Although this book deals with the theoretical aspects of electronic transport, a "minimal" knowledge of the experimental setup for AC transport is always an advantage in understanding different aspects of the problem.

DC transport properties can be directly measured using four-probe techniques, but measurement of transport properties at finite frequencies is quite intricate. Low frequency conductivity (at radio frequencies) can be directly measured but at infrared (IR) and optical frequencies, the optical/IR measurement techniques such as the ellipsometry are used. A basic set-up to measure reflectivity/absorptivity of a sample is given below. From the experimental reflectivity/absorptivity the conductivity can be computed using the Kramer-Kronig analysis as explained below. In an experiment, a beam of photons from a monochromatic source of light is made to incident on a sample using the ellipsoidal rotating mirror which further transmits or reflects the beam (Figure 1.4). In one orientation of the mirror (pointing toward the sample), photons fall on the sample and reflected photons are captured by a detector (Figure 1.4). In another orientation of the mirror (mirror pointing toward the detector), photons directly fall on the detector. With this, one obtains the ratio of the number of photons reflected by the sample to the total number of photons incident on the sample at a particular frequency. This ratio is called reflectance and it is a function of frequency. The reflectance ($R(\omega)$) can be connected with microscopic transport quantities:

$$R(\omega) = \frac{(n(\omega) - 1)^2 + k(\omega)^2}{(n(\omega) + 1)^2 + k(\omega)^2}. \tag{1.20}$$

Here $n(\omega)$ and $k(\omega)$ are the real and imaginary parts of the refractive index ($N(\omega)$) of the material. In the following discussion we will derive this result.

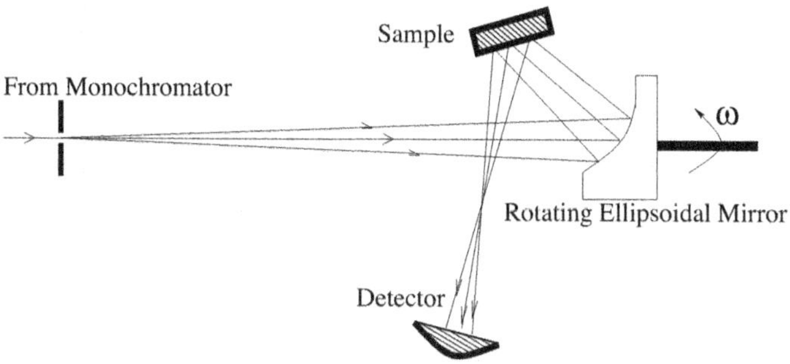

FIGURE 1.4: A sketch of an apparatus for measuring reflectivity/absorptivity of a material.

The Maxwell equations (in Gaussian units) for an isotropic material can be written as:

$$\nabla.\mathbf{E} = 0. \tag{1.21}$$

$$\nabla \times \mathbf{E} = \frac{-\mu}{c}\frac{\partial \mathbf{H}}{\partial t}. \tag{1.22}$$

$$\nabla \times \mathbf{H} = \frac{\epsilon}{c}\frac{\partial \mathbf{E}}{\partial t} + \frac{4\pi}{c}\sigma\mathbf{E}. \tag{1.23}$$

$$\nabla.\mathbf{H} = 0. \tag{1.24}$$

Here μ, σ, and ϵ is the permeability, the conductivity, and the dielectric function of a material. Taking curl on both sides of Equation (1.22) leads to

$$\nabla \times (\nabla \times \mathbf{E}) = \frac{-\mu}{c}\left[\frac{\partial}{\partial t}(\nabla \times \mathbf{H})\right] \tag{1.25}$$

Using the vector identity $\nabla \times (\nabla \times \mathbf{E}) = \nabla(\nabla.\mathbf{E}) - \nabla^2\mathbf{E}$ and above Maxwell's equations (Equations (1.21) and (1.23)), Equation (1.25) becomes:

$$\nabla^2\mathbf{E} = \frac{\mu\epsilon}{c^2}\frac{\partial^2 \mathbf{E}}{\partial t^2} + \frac{4\pi\mu}{c^2}\sigma\frac{\partial \mathbf{E}}{\partial t} \tag{1.26}$$

In experiments the optical properties are generally measured using plane polarized monochromatic light. Thus, consider the electric field to be

$$\mathbf{E} = \mathbf{E}_0 \exp\left[i(\mathbf{q}.\mathbf{r} - \omega t)\right] = \mathbf{E}_0 \exp\left[i(qz - \omega t)\right]. \tag{1.27}$$

where $\mathbf{E}_0$ is perpendicular to the propagation direction, i.e., in the direction of the wave vector $\mathbf{q}$ which is along the direction of the $z-$axis (say). Let the $z-$axis be perpendicular to the plane surface of the material from which the reflection is measured (this surface of the material is in the $xy-$plane). In Equation (1.27) ω is the frequency and q is the wave vector of the electric field. Substitution of Equation (1.27) into (1.26), leads to:

$$q^2 = \left(\frac{\mu\omega^2}{c^2}\right)\left[\epsilon + i\frac{4\pi}{\omega}\sigma\right]. \tag{1.28}$$

Notice that the wave vector acquires an imaginary part due to finite conductivity of the material which leads to the damping of the field inside the material. The refractive index N of a material is defined as c/v where v is the magnitude of the velocity of light inside the material. As $\omega = qv$, the magnitude of the wave vector can be written as

$$q = \frac{\omega}{c}N = \frac{\omega}{c}(n + ik), \tag{1.29}$$

where the real (n) and the imaginary part (k) of the refractive index are introduced. Using this, the electric field inside the material can be rewritten as:

$$\mathbf{E} = \mathbf{E}_0\left[\exp\left(-\frac{\omega}{c}kz\right)\right]\exp i\left(\frac{\omega}{c}nz - \omega t\right). \tag{1.30}$$

The first exponential in the above equation shows that the field intensity is exponentially damped inside the material. This is due to the dissipative processes. To quantitatively capture this, one defines the absorption coefficient:

$$\alpha = \frac{-1}{I}\frac{dI}{dz}, \tag{1.31}$$

where I is the intensity of the field. Since intensity is proportional to the square of the amplitude of the electric field $I = \mathbf{E}_0^2 \exp\left[-2k\frac{\omega}{c}z\right]$ one obtains

$$\alpha = \frac{2k\omega}{c}. \tag{1.32}$$

The absorption coefficient is an important parameter of a material. In the following, its frequency dependence is discussed for a material that obeys the

Drude model. To express dielectric function and conductivity in terms of n and k, square both sides of Equation (1.29) and compare with Equation (1.28), one will get

$$\epsilon = \frac{n^2 - k^2}{\mu}. \tag{1.33}$$

$$\sigma = \frac{nk\omega}{2\pi\mu}. \tag{1.34}$$

The dielectric function ϵ and the conductivity σ in the above equations are real quantities (as evident, right hand sides (RHSs) of both equations contain only real quantities and no i). Since these are response functions, due to casualty, they must have the imaginary parts too (this will be made evident in the Kubo formalism in Chapter 3). Thus, we need to redefine our notation. The above ϵ and σ will be replaced by ϵ_1 and σ_1. And the total (complex) dielectric function will be defined as:

$$\epsilon = \frac{N^2}{\mu} = \epsilon_1 + i\epsilon_2. \tag{1.35}$$

Using the expression (1.29) and (1.34), the real and imaginary parts of the dielectric function are

$$\epsilon_1 = \frac{n^2 - k^2}{\mu} \qquad \epsilon_2 = \frac{2nk}{\mu} = \frac{4\pi\sigma_1}{\omega}. \tag{1.36}$$

Thus, if n and k can be experimentally measured, the real and imaginary parts of the dielectric function and of the conductivity can be extracted (from the above equations). σ_2 can be extracted from σ_1 by Kramers-Kronig (KK) transformation. Our next task is to connect n and k with the reflectivity or absorptivity of the material which is what experiments measure. Define the transmitted, reflected, and incident electric field as $\mathbf{E}_t$, $\mathbf{E}_r$, and $\mathbf{E}_i$, respectively. Similarly define the above components for the magnetic field. From the boundary conditions:

$$\mathbf{E}_i + \mathbf{E}_r = \mathbf{E}_t. \tag{1.37}$$

$$\mathbf{H}_i - \mathbf{H}_r = \mathbf{H}_t. \tag{1.38}$$

Remember that all the components of the electric and magnetic fields are perpendicular to the $z-$axis (the direction of propagation) and are in a plane which is parallel to the plane surface of the material (on which the reflection is measured). The plane polarized electric and magnetic fields are given as $\mathbf{E} = \mathbf{E}_0 e^{i(\omega/c)Nz-i\omega t}$ and $\mathbf{H} = \mathbf{H}_0 e^{i(\omega/c)Nz-i\omega t}$. From the Maxwell equation (1.22), for a nonmagnetic material ($\mu = 1$), say, it is easy to prove that $\mathbf{H} = N\mathbf{E}$ inside the material, and $\mathbf{H} = \mathbf{E}$ outside the material ($N = 1$ for vacuum). Therefore,

$$\mathbf{E}_i + \mathbf{E}_r = \mathbf{E}_t. \tag{1.39}$$

$$\mathbf{E}_i - \mathbf{E}_r = N\mathbf{E}_t. \tag{1.40}$$

Solving the above:

$$\mathbf{E}_i = \frac{1}{2}(1 + N)\mathbf{E}_t \tag{1.41}$$

$$\mathbf{E}_r = \frac{1}{2}(1 - N)\mathbf{E}_t \tag{1.42}$$

The reflectivity is defined as the ratio of the square of the reflected part of the electric field to the square of the incident part:

$$R = \frac{|E_r|^2}{|E_i|^2} = \frac{|1 - N|^2}{|1 + N|^2} = \frac{|1 - n - ik|^2}{|1 + n + ik|^2} = \frac{(1 - n)^2 + k^2}{(1 + n)^2 + k^2}. \tag{1.43}$$

And the absorptivity is:

$$A = 1 - R = \frac{4n}{(1 + n)^2 + k^2}. \tag{1.44}$$

In a typical experiment, the reflectivity of a material is measured at various frequencies, and by Kramers-Kronig analysis the n and k can be extracted at those frequencies. This inversion involves an extensive numerical procedure [7]. From the knowledge of the real and imaginary parts of the refractive index, the complex conductivity and the complex dielectric function can be determined. For example, for a nonmagnetic material ($\mu = 1$) one has from

$n^2 - k^2 = \epsilon_1$ and $2nk = \epsilon_2$:

$$n = \frac{1}{\sqrt{2}}\sqrt{\sqrt{\epsilon_1^2 + \epsilon_2^2} + \epsilon_1}$$
$$k = \frac{1}{\sqrt{2}}\sqrt{\sqrt{\epsilon_1^2 + \epsilon_2^2} - \epsilon_1} \qquad (1.45)$$

1.3 The Drude model at finite frequencies for simple metals

In this section we will compute the reflectivity ($R(\omega)$) of a "simple metal" (like gold or silver) using the Drude model at finite frequencies, and we will analyze the expressions in various frequency regimes.

We have learned that in the Drude model electrons in a metal are treated as classical particles (much like gas molecules) bumping with the stationary ion cores. Now we adopt a slightly different approach. Under the action of an external electric field, the equation of motion for an electron can be written as:

$$m\frac{d^2\mathbf{r}}{dt^2} = -e\mathbf{E}(t) - m\frac{1}{\tau_{Drude}}\frac{d\mathbf{r}}{dt}, \qquad (1.46)$$

which is nothing but Newton's second law of motion. Here $\frac{1}{\tau_{Drude}}$ is the average rate at which the momentum of an electron is being degraded due to collision processes with the ion cores. We will notice that τ_{Drude} is equal to ATFA, i.e., τ. Let us suppose that the electric field $\mathbf{E}(t)$ varies as $\mathbf{E}_0\cos(\omega t)$. With this, the above equation of motion can be solved for momentum $\mathbf{p} = md\mathbf{r}/dt$. It transforms to a first order differential equation in momentum with solution:

$$\mathbf{p}(t) = \underbrace{-e\frac{\tau_{Drude}}{1+\tau_{Drude}^2\omega^2}\mathbf{E}_0\cos(\omega t)}_{In\ phase\ response\ (dissipative)} - \underbrace{e\frac{\omega\tau_{Drude}^2}{1+\tau_{Drude}^2\omega^2}\mathbf{E}_0\sin(\omega t)}_{Out-of-phase\ response\ (reactive)} \quad . \qquad (1.47)$$

It has two terms: (1) the in-phase response which goes like $\cos(\omega t)$ and leads to dissipation, and (2) out-of-phase response which goes like $\sin(\omega t)$ and does not lead to dissipation. The current density induced due to the application of the electric filed is given by

$$\begin{aligned} \mathbf{j}(t) &= -ne\mathbf{p}(t)/m \\ &= e^2(n/m)\frac{\tau_{Drude}}{1+\tau_{Drude}^2\omega^2}\mathbf{E}_0\cos(\omega t) + e^2(n/m)\frac{\omega\tau_{Drude}^2}{1+\tau_{Drude}^2\omega^2}\mathbf{E}_0\sin(\omega t). \end{aligned} \tag{1.48}$$

The average heat dissipated per unit volume (Joule heating) in the material in one cycle of the applied field is given by $\frac{\omega}{2\pi}\int_0^{\frac{2\pi}{\omega}} dt\mathbf{j}(t).\mathbf{E}(t)$. Clearly, it is only due to the dissipative component not due to the reactive component, as the product $\sin(\omega t)\cos(\omega t)$ vanishes when averaged over the period[§]. Thus,the dissipative part of the conductivity is given by

$$\sigma_1(\omega) = \sigma_{DC}\frac{1}{1+\tau_{Drude}^2\omega^2}. \tag{1.49}$$

And the reactive part of the conductivity is given by

$$\sigma_2(\omega) = \sigma_{DC}\frac{\tau_{Drude}\omega}{1+\tau_{Drude}^2\omega^2}. \tag{1.50}$$

Here the DC conductivity is defined as $\sigma_{DC} = \frac{ne^2\tau_{Drude}}{m}$. We notice that $\tau_{Drude} = \tau$. In the following we do not distinguish between the two. The σ_1 and σ_2 are also called the real and imaginary parts of the conductivity. From Equation (1.36) the imaginary part of the dielectric function is given by $\frac{4\pi}{\omega}\sigma_1$. The real part of it can be computed either by using Kramers-Kronig relation or by the following method:

Integrating Equation (1.47) over time leads to

$$\mathbf{r}(t) = \underbrace{-(e/m\omega)\frac{\tau}{1+\tau^2\omega^2}\mathbf{E}_0\sin(\omega t)}_{reactive} + \underbrace{(e/m\omega)\frac{\omega\tau^2}{1+\tau^2\omega^2}\mathbf{E}_0\cos(\omega t)}_{dissipative}. \tag{1.51}$$

The integration constant is chosen in such a way that in the remote past ($t \to -\infty$) when the field is zero, $\mathbf{r}(-\infty)$ vanishes and the field is adiabatically switched on. Let us assume that for $\mathbf{r}(-\infty) = 0$ all the negative charges overlap

§Refer to problem 1 in the last chapter.

with positive charges. The net induced polarization per unit volume at any time is given as $\mathscr{P} = -en\mathbf{r}$. The displacement is $\mathbf{D} = \epsilon\mathbf{E} = \mathbf{E} + 4\pi\mathscr{P}$. Thus, ϵ has the following components:

$$\epsilon_1 = 1 - \omega_p^2 \frac{\tau^2}{1+\tau^2\omega^2}. \tag{1.52}$$

$$\epsilon_2 = \omega_p^2 \frac{\tau}{\omega(1+\tau^2\omega^2)}. \tag{1.53}$$

Where ω_p is the plasma frequency $= \sqrt{\frac{4\pi ne^2}{m}}$. From ϵ_1 and ϵ_2 the real and imaginary parts of the refractive index (Equation (1.45)) can be computed and the reflectance and the absorption coefficient can be analyzed. Below we analyze these in various regimes of interest.

Case I: When $\omega << \frac{1}{\tau} << \omega_p$ (low frequency regime)

In the leading order, the dielectric functions (Equations (1.52) and (1.53)) can be simplified to:

$$\epsilon_1 \simeq -\omega_p^2\tau^2, \qquad \epsilon_2 \simeq \frac{\omega_p^2}{\omega}\tau$$

$\epsilon_1^2 + \epsilon_2^2 \approx \epsilon_2^2$ (as $\epsilon_2 \gg \epsilon_1$). Hence,

$$n = k \simeq \omega_p\sqrt{\frac{\tau}{2\omega}}, \qquad \text{and,} \qquad A(\omega) \simeq \frac{2}{\omega_p}\sqrt{\frac{2\omega}{\tau}} << 1. \tag{1.54}$$

And the reflectance $R = 1 - A \simeq 1$. *Thus at low frequencies metals are good reflectors.* This is called the Hagen-Rubens regime. The absorption coefficient is

$$\alpha = \frac{2k\omega}{c} = \frac{2}{c}\omega\omega_p\sqrt{\frac{\tau}{2\omega}} = \frac{2}{c}\omega_p\sqrt{\frac{\omega\tau}{2}} \approx 1, \tag{1.55}$$

which is small and varies as $\sqrt{\omega}$. Also, the penetration depth, which is reciprocal of the absorption coefficient, varies as:

$$\delta_{classical} = \frac{c}{k\omega} = \frac{c}{\omega_p}\sqrt{\frac{2}{\omega\tau}} \propto \frac{1}{\sqrt{\omega}}. \tag{1.56}$$

The effect where $\delta \gg l$ ($l = v\tau$ is the electron mean free path) is known as the normal skin effect. But if $l \gtrsim \delta_{classical}$, then the effect is called the anomalous skin effect. By Pippard's ineffectiveness argument, or by more sophisticated calculations it can be shown that $\delta_{anomalous} \sim \frac{1}{\omega^{1/3}}$ (see for details [8]).

Case II: When $\frac{1}{\tau} << \omega << \omega_p$ (intermediate frequency regime)

In the leading order the dielectric function can be simplified to:

$$\epsilon_1 \simeq -\frac{\omega_p^2}{\omega^2}, \qquad \epsilon_2 \simeq \frac{1}{\tau\omega}\left(\frac{\omega_p}{\omega}\right)^2. \tag{1.57}$$

In this case, ϵ_1 is dominant over ϵ_2. Thus $n \simeq \sqrt{\epsilon_1}$, and $k \simeq 0$. Therefore, the absorptivity varies as:

$$A(\omega) \simeq \frac{4\omega}{\omega_p} \ll 1. \tag{1.58}$$

Hence, $R(\omega) = 1 - A \simeq 1$. The metal is still reflecting. The absorption coefficient $\alpha(\omega)$ in this case is $\propto \frac{1}{\omega}$. This is called the relaxation regime.

Case III: When $\frac{1}{\tau} << \omega_p << \omega$ (high frequency regime)

Again in the leading order, the dielectric function will be:

$$\epsilon_1 \simeq 1 (\text{in the next order it is } 1 - \frac{\omega_p^2}{\omega^2}), \qquad \epsilon_2 \simeq \frac{1}{\tau}\frac{\omega_p^2}{\omega^3} << 1. \tag{1.59}$$

ϵ_1 again dominates in this regime and the absorptivity $A(\omega) \simeq 1$ and reflectivity $R(\omega) \simeq 0$. And the absorption coefficient $\alpha(\omega) \sim \frac{1}{\omega^2}$. The metal in this regime is largely transmitting (e.g., metals are largely transparent to UV radiation). All these three regimes are summarized in Table (1.1).

Thus, we notice that the AC Drude model *qualitatively* captures the optical/IR properties of simple metals. One must recognize that it only crudely captures the optical features of simple metals. According to it all metals are good reflectors at intermediate frequency (refer to Table 1.1) and given that the visible band is narrow (refer to the electromagnetic spectrum), all the simple metals should appear white to the human eye. This is in contradiction to observation, for example, copper appears yellowish, gold and silver have their own color and luster, etc. To account for these optical properties, one has to go beyond the simple Drude model and invoke the band structure and the location of the Fermi level for a given material. We will discuss these points in a subsequent section.

TABLE 1.1: Reflectance and absorption coefficient for simple metals that obey the Drude model

τ is the Drude scattering rate and $\omega_p = \sqrt{\frac{4\pi ne^2}{m}}$ is the plasma frequency.		
Case I: $\omega << \frac{1}{\tau} << \omega_p$	Case II: $\frac{1}{\tau} << \omega << \omega_p$	Case III: $\frac{1}{\tau} << \omega_p << \omega$
$R(\omega) \simeq 1 - \frac{2}{\omega_p}\sqrt{\frac{2\omega}{\tau}} \simeq 1$	$R(\omega) \simeq 1 - \frac{4\omega}{\omega_p} \simeq 1$	$R(\omega) \simeq 0$
$\alpha(\omega) \simeq \frac{2}{c}\omega_p\sqrt{\frac{\omega\tau}{2}} \sim \sqrt{\omega}$	$\alpha(\omega) \sim \frac{1}{\omega}$	$\alpha(\omega) \sim \frac{1}{\omega^2}$
REFLECTING(HAGEN-RUBENS)	REFLECTING (RELAXATION)	UV TRANSPARENT

1.4 The Lorentz model for simple insulators

We have considered the Drude model which is applicable to metals (in which free conduction electrons scatter off the localized ion cores). In an insulator we do not have free conduction electrons, instead the electrons are bounded to their parent atoms. The Lorentz oscillator model deals with an insulator. Let us visualize a toy model for an insulator (like diamond). Consider all the electrons are bound. For simplicity, consider that each atom has one bound electron and there is no interaction between electrons of separate atoms (atoms do not interact with each other, i.e., the dilute limit; again highly unrealistic), and in the dilute limit the field seen by an electron is equal to the external applied field, as the induced field by the neighboring atoms is weak and can be neglected¶. We introduce a classical model of an

¶But in real insulators these local field corrections are important, consult for a detailed account [7].

atom in which an electron having charge $-e$ and mass m is bounded with the nucleus by a harmonic force and whose motion in the presence of an external electric field is described by the following equation:

$$m\frac{d^2\mathbf{r}}{dt^2} + m\mathbf{r}\omega_0^2 = -e\mathbf{E} - m\Gamma\frac{d\mathbf{r}}{dt}. \tag{1.60}$$

Here, the term $m\Gamma\frac{d\mathbf{r}}{dt}$ is the damping force and $m\mathbf{r}\omega_0^2$ is the restoring force. Let us suppose that the electric field $\mathbf{E}$ and distance $\mathbf{r}$ of the electron from the equilibrium position varies as[‖]:

$$\mathbf{E}(t) = \mathbf{E}_0 \exp(-i\omega t) \tag{1.61}$$

$$\mathbf{r}(t) = \mathbf{r}_0 \exp(-i\omega t). \tag{1.62}$$

On substituting these equations into Equation (1.60), we have

$$\mathbf{r}_0 = -\frac{e}{m}\mathbf{E}_0\frac{1}{\omega_0^2 - \omega^2 - i\omega\Gamma}. \tag{1.63}$$

The electric field $\mathbf{E}$ acting on the electron will induce a dipole moment whose time independent part is

$$\mathbf{p}_0 = -e\mathbf{r}_0 \tag{1.64}$$

$$\mathbf{p}_0 = \frac{e^2}{m}\mathbf{E}_0\frac{1}{\omega_0^2 - \omega^2 - i\omega\Gamma}. \tag{1.65}$$

The induced dipole moment is related to the applied electric field as:

$$\mathbf{p} = \alpha(\omega)\mathbf{E}, \tag{1.66}$$

where $\alpha(\omega)$ is the atomic polarizability. On comparing Equations (1.65) and (1.66), we get

$$\alpha(\omega) = \frac{e^2}{m}\frac{1}{\omega_0^2 - \omega^2 - i\omega\Gamma}. \tag{1.67}$$

If the system has n atoms per unit volume, then total polarizability per unit volume is

$$\mathcal{P}(\omega) = n\alpha(\omega). \tag{1.68}$$

[‖]One can assume a real electric field like $E_0\cos(\omega t)$. Then, as we have done with the Drude model in the previous section, one will have a dissipative response and reactive response. Both treatments give the same result.

The displacement $\mathbf{D}$ is given by $\mathbf{E} + 4\pi\mathcal{P}$ or by $\epsilon\mathbf{E}$. Thus, the complex dielectric function is:

$$\epsilon = 1 + \frac{\omega_p^2}{\omega_0^2 - \omega^2 - i\omega\Gamma}. \tag{1.69}$$

With this, the real and the imaginary parts of the dielectric function take the form:

$$\epsilon_1 = 1 + \frac{\omega_p^2(\omega_0^2 - \omega^2)}{(\omega_0^2 - \omega^2)^2 + \omega^2\Gamma^2}. \tag{1.70}$$

$$\epsilon_2 = \frac{\omega_p^2\Gamma\omega}{(\omega_0^2 - \omega^2)^2 + \omega^2\Gamma^2}. \tag{1.71}$$

Based on this dielectric function for the Lorentz model we will analyze the optical properties of an ideal insulator. Although the Lorentz oscillator model for an insulator is a very simplified model and is not quantitatively applicable to real life insulators, it captures some important qualitative features of typical insulators. A brief discussion of the optical properties of real insulators and real metals is given in the next section.

To study various important frequency regimes of interest, let us assume that $\Gamma << \omega_0 \sim \omega_p$ which is a reasonable assumption for real insulators. Consider the following regimes:

Case I: Low frequency regime: When $\omega \simeq \Gamma << \omega_0$

In the leading order, the dielectric functions can be simplified to:

$$\epsilon_1 \simeq 1 + \frac{\omega_p^2}{\omega_0^2} \simeq 2, \qquad \epsilon_2 \simeq \frac{\omega_p^2\Gamma\omega}{\omega^4} = (\frac{\omega_p}{\omega_0})^2(\frac{\Gamma}{\omega_0})(\frac{\omega}{\omega_0}) << 1.$$

Therefore,

$$n \simeq \sqrt{\epsilon_1} \simeq 1.4, \qquad \text{and,} \qquad k \simeq 0. \tag{1.72}$$

And the reflectance $R = \frac{(1-n)^2+k^2}{(1+n)^2+k^2} \simeq 0.03$ is very weak. Thus, at low frequencies, insulators are bad reflectors (compare this with the Hagen-Rubens regime of metals). The absorption coefficient is

$$\alpha = \frac{2k\omega}{c} \simeq 0. \tag{1.73}$$

Thus, small absorption coefficient and weak reflectance implies that the insulators in this regime are good transmitters. In other words, they are transparent to low frequency radiation, consider for example, crystals of $NaCl$ (ordinary salt crystals) or KCl crystals or ordinary silica glass, etc.

Case II: Intermediate frequency regime: When $\omega_0 \simeq \omega$

Assume that: $|\omega_0 - \omega| = \delta << \Gamma,\ \omega,\ \omega_0$ (δ is the smallest parameter). In the leading order the dielectric function can be simplified to:

$$\epsilon_1 \simeq 1 + \frac{2\omega_p^2\omega_0\delta}{(2\omega_0\delta)^2 + \Gamma^2\omega_0^2} \simeq 1 + 2(\frac{\omega_p}{\Gamma})(\frac{\omega_p}{\omega_0})(\frac{\delta}{\Gamma}) = O(1)(\text{order of one}), \quad (1.74)$$

$$\epsilon_2 \simeq (\frac{\omega_p}{\omega_0})^2(\frac{\omega}{\Gamma}) >> 1. \quad (1.75)$$

In this case, ϵ_2 is dominant over ϵ_1. Thus, $n \simeq k \simeq \sqrt{\epsilon_2/2} >> 1$. Therefore, the reflectivity $R \sim 1$. The absorption coefficient $\alpha(\omega) = 2k\omega/c$ is appreciable and varies as $\frac{\omega}{c}\frac{\omega_p}{\omega_0}\sqrt{\frac{\omega}{\Gamma}} \sim \omega^{3/2}$. Thus the insulator starts to reflect but there is appreciable damping inside the insulator. This is called the relaxation regime.

Case III: High frequency regime: When $\omega_0 << \omega$

Again in the leading order, the dielectric function will be:

$$\epsilon_1 \simeq 1(\text{in the next order it is } 1 - \frac{\omega_p^2}{\omega^2}), \qquad \epsilon_2 \simeq (\frac{\omega_p}{\omega})^2\frac{\Gamma}{\omega} << 1. \quad (1.76)$$

ϵ_1 again dominates in this regime. As $n \simeq 1$ and $k \simeq 0$, the reflectivity is again very weak $R \simeq 0$. And the absorption coefficient which is proportional to k is also negligible. Thus, the insulator again turns transparent (called UV transparency like in metals) and is not dissipating. All these three regimes are summarized in Table (1.2).

TABLE 1.2: Reflectance and absorption coefficient for simple insulators that obey the Lorentz model. Γ is the Lorentz scattering rate, ω_p is the plasma frequency, and ω_0 is the harmonic frequency.

Case I: Low frequency regime: When $\omega \simeq \Gamma << \omega_0$	Case II: Intermediate frequency regime: When $\omega_0 \simeq \omega$	Case III: High frequency regime: When $\omega_0 << \omega$
$R(\omega) << 1$	$R(\omega) \simeq 1$	$R(\omega) << 1$
$\alpha(\omega) \simeq 0$	$\alpha(\omega) \sim \omega^{3/2}$ (appreciable)	$\alpha(\omega) \simeq 0$
INSULATOR APPEARS TRANSPARENT	REFLECTING (BUT WITH INTERNAL DISSIPATION)	UV TRANSPARENT

1.5 Brief discussion of the optical properties of real metals and real insulators

1.5.1 Real metals

We observe that the simple Drude model can qualitatively explain the optical properties of metals. The model predicts that, in the low frequency regime ($\omega << \omega_p$), metals are good reflectors, and when the frequency of incident radiation becomes of the order of plasma frequency or greater than that, they turn into good transmitters (UV transparency). The plasma frequency is several electron volts in metals. Thus, according to the Drude model metals are good reflectors in the visible spectrum ($700\ nm \simeq 0.3\ eV < \omega < 400\ nm \simeq 0.5\ eV$). If a material reflects all wavelengths in a uniform way in the visible spectrum, then the material must appear white to the human eye. Thus, silver appears whitish. However, copper and gold have their reddish or "copperish" and "goldish" luster. If metals uniformly reflect in the visible spectrum, then they all should appear whitish in color. Therefore, it implies that within the Drude model it is not possible to address the individual differences between various metals.

To take full account of the reflecting properties of metals, one must go beyond the Drude model and one must invoke the electronic band structure of metals. Band theory classifies metals as materials with partially filled conduction band(s). The Fermi level passes through the conduction band and the electrons from the filled levels below the Fermi level can be excited to the levels above it; even with vanishingly small energies as there is no energy gap. This *intra-band* electronic conduction is included in the Drude model through a relaxation time. This works well for DC transport and very low frequency electromagnetic radiation like far-IR or IR (Infra-Red).

As the incident frequency is increased, it becomes possible to excite electrons from valence bands or from filled states below the Fermi level. This mechanism is called the *inter-band* conduction, and it is not included in the Drude model. Thus, to fully describe the electromagnetic response of metals over an extended frequency regime one has to use a combination of the Drude model and Lorentz model. In the experiments, the reflectance or conductivity is fitted with a combination of Drude model and the Lorentz model (called jointly the Drude-Lorentz model). The model parameters are determined from the best fits [9]. Thus, one is able to qualitatively address the electromagnetic response over an extended frequency regime. However, there are other complications arising from electron-electron and electron-phonon interactions that modify the electromagnetic response in a subtle way. For example, at low temperatures, metals show extra absorption features in the IR spectra due to the second order processes in which an incident photon is absorbed by metal and it generates an electron hole pair along with an acoustic phonon. This mechanism is called the Holstein mechanism [10]. These subtle effects are not

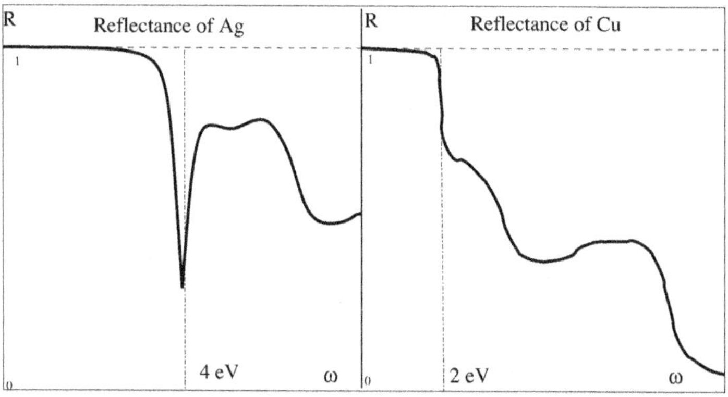

FIGURE 1.5: Schematic depiction of the reflectance spectra of Ag and Cu. Vertical dot-dashed lines show the regimes where d-band transitions are induced.

included in the simple Drude-Lorentz model. One must use more sophisticated fully quantum approaches to address these issues, for example, the memory function formalism (Chapter 4) or the Holstein-Boltzmann equation [10].

Returning back to the qualitative picture, one can understand the differences in the appearances of, for example, silver (Ag) and copper (Cu) by invoking the band theory. Figure 1.5 schematically depicts the reflectance spectrum of Ag and Cu. In the case of Ag, the reflectance is roughly $\simeq 1$ up to 4 eV (remember 4 eV is well into UV). After that, it falls drastically for a short interval of frequencies, and again it raises to a large value in the intermediate range from 4 eV to 8 eV, and after that it falls again. These peculiar features of reflectance in noble metals are due to the presence of d-bands below the Fermi level. Thus, the fall in R around 4 eV in Ag is due to the onset of the inter-band absorption where electrons are excited from the d-band below the Fermi level to the conduction band above the Fermi level. The visible spectrum is roughly from one-tenth of an electron volt to less than one electron volt. In this interval reflectance of Ag is roughly constant and is near to one. Thus, Ag reflects all the frequencies in the visible spectrum and appears whitish.

In the case of Cu, the d-band lies only 2 eV below the Fermi level. Thus, the shoulder in reflectance (Figure 1.5) shifts to lower frequencies and appears around 2 eV. If the spectrum is inspected closely, it is seen that in the case of Cu the reflectance starts to become less than that of Ag from 700 $nm \simeq$ 0.3 eV onward and slowly decreases up to 2 eV. At 2 eV, it shows a sharp reduction feature, that is the shoulder (Figure 1.5). The 700 nm wavelength corresponds to the red color. The color of a substance is determined by the addition of corresponding colors of reflected waves by it. Due to this peculiarity Cu appears reddish or shows "copperish" luster. Thus, within the band theory, many direct observational features can be understood. For more details, refer to [7].

In conclusion, we notice that real metals are much more complex than those pictured in the Drude model and one has to consider electronic band structure to understand the electromagnetic response of metals. At very low temperatures and low frequencies electron-phonon and/or electron-electron interaction effects may modify the optical/IR spectra [10].

1.5.2 Real insulators

In the Lorentz model, an insulator is visualized as a material with a collection of atoms with bound electrons. The electrons are not free and under the influence of an external field they oscillate about their equilibrium positions. Since the Lorentz model is based on the classical picture, one must keep the fact in mind that the correct way to describe electrons in atoms and their interaction with light is through proper quantum treatment. Nevertheless, the very crude classical picture of Lorentz can explain some of the observed features in a qualitative way. The oscillations of the bound electrons are treated in the harmonic approximation, and the equation of motion was written down with a damping term to account for dissipation. The characteristic frequency scale of the system is given by the natural frequency $\omega_0 = \sqrt{\frac{k}{m}}$ where m is the mass of the electron and k is the "spring constant" of the bonds. From the analysis of the Lorentz model we observed that in the low frequency regime ($\omega << \omega_0$) the reflectance of an insulator is very small and at the same time internal dissipation is also small. Thus, the insulator must appear transparent in this frequency regime. Indeed, KCl crystals, for example, are transparent to the human eye (as the frequency range of the visible regime is small as compared to ω_0).

In the intermediate frequency regime, when $\omega \sim \omega_0$, the insulator turns into a good reflector and also has appreciable internal dissipation (the absorption coefficient is appreciable in this regime). In the high frequency regime, when $\omega >> \omega_0$, the bound electrons appear approximately free to such high frequency radiation and the reflectance falls down like in the case of metals.

It turns out that the characteristic frequency ω_0 or energy $\hbar\omega_0$ corresponds to the energy required to start inter-band transitions. It can be seen with a quantum treatment of the Lorentz model [7]. The optical properties are drastically affected by the size of the inter-band gap or band gap. Let us consider an interesting case of *Ge* or *Si* and KCl. They are band insulators at zero temperature. In *Ge* and *Si* the band gaps are smaller (roughly 1 eV^{**}). Due to this smaller gap, even at room temperatures, there is a sizable number of electrons in the conduction band (due to thermal excitations). Thus, *Si* and *Ge* show metallic reflectance at optical wavelengths at room temperatures (that comes from conduction electrons). They are not transparent in contrast to KCl. In KCl the band gap is very large, of the order of 7 eV. Thus, there are no thermally excited carriers in the conduction band at room temperature, and metallic reflectance is absent. Therefore, KCl crystals are transparent at optical wavelengths, in accord with the Lorentz model. These are very basic inferences but nevertheless give a reasonable understanding of simple insulators. For more quantitative discussion refer to [7].

**Si band gap is ~ 1.1 eV, and the Ge band gap is ~ 0.8 eV.

1.6 Summary points

1. The first notable theoretical investigation into the nature of electrical conduction in metals is due to Paul Drude. The theory apparently was able to address the Wiedemann-Franz law but failed, including other problems, to address temperature dependence of electrical and thermal conductivities. The founding fathers such as Lorentz and his contemporaries tried hard to resolve the issues but it yielded to Sommerfeld and Bloch when quantum mechanics was in place.
2. We learned that optical properties of metals and insulators can be qualitatively understood within the classical Drude-Lorentz theory.
3. We reviewed some of the complications that arise when one considers realistic metals and insulators. The electronic band structure effects, which are neglected in the Drude theory, have to be taken into account. We are able to understand why copper appears reddish while silver appears whitish. Similarly, the differences between opaque Si or Ge with transparent KCl can be understood just by considering the magnitude of the band gap.

Bibliography

[1] A. H. Wilson, *The Theory of Metals*, CUP (1953).

[2] N. Ashcroft and N. Mermin, *Solid State Physics*, Brooks-Cole (1976).

[3] P. B. Allen and W. H. Butler, *Physics Today*, December issue, 44-49 (1978).

[4] P. B. Allen et al., *Phys. Rev. B*, **34**, 4331 (1986).

[5] R. E. Peierls, *Quantum Theory of Solids*, Oxford University Press (1955).

[6] L. P. Kadanoff and G. Baym, *Quantum Statistical Mechanics*, Perseus Books (1994); R. E. Prange and L. P. Kadanoff, *Phys. Rev.* **134**, A566 (1964); W. Haensch and G. D. Mahan, *Phys. Rev. B* **28**, 1902 (1983); R. Kubo, *Rep. Prog. Phys.*, **29**, 255 (1966); R. Zwanzig, *Nonequilibrium Statistical Mechanics*, Oxford University Press (2001).

[7] F. Wooten, *Optical Properties of Solids*, Academic Press (1972).

[8] J. M. Ziman, *Principles of the Theory of Solids*, CUP (1979).

[9] H. R. Philipp and H. Ehrenreich, *Phys. Rev.* **131**, 2016 (1963).

[10] T. Holstein, *Phys. Rev.* **96**, 535 (1954); T. Hosltein, *Ann. Phys. (N. Y.)* **29**, 410 (1964).

Chapter 2

The traditional Boltzmann equation based approaches

Gas-like quasi-particles in a solid are in fact very complex objects not understandable by intuition. Thus a Boltzmann equation approach to study physics of a given solid is subject to very basic questions which would not be asked when we work on a Boltzmann equation for a gas.

R. Kubo*

From the historical survey in the previous chapter, we learned that the problems faced by the Drude-Lorentz theory were partly rectified by the Bloch-Boltzmann theory. This chapter is devoted to the Bloch-Boltzmann theory and more generally to the application of the Boltzmann equation to electrons in metals. In the first chapter we also learned that it is possible to deduce transport quantities like the real and imaginary parts of the conductivity from experimental reflectance spectrum of a material. The transport quantities have hidden information regarding microscopic interactions and scattering mechanisms of quasiparticles. The first model–also the most rudimentary one–is the Drude model which we have discussed in the previous chapter in the context of simple metals. We have also discussed the Lorentz model which is a rudimentary model for an insulator.

This chapter deals with a general treatment of transport and transport coefficients. We are going to discuss the semiclassical model of the electron

---

**Acta Physica Austriaca, Suppl. X*, Pages 301-340 (1973).

dynamics. This provides the foundation for the application of the Boltzmann equation to transport in metals. We introduce Chamber's formulation of the Boltzmann equation which is a more transparent approach and is equivalent to the relaxation time approximation. The cases of DC and AC transport are discussed in the context of electrical conduction in metals. Then Boltzmann equation beyond relaxation time approximation is introduced. The traditional theory of the Bloch-Boltzmann equation is introduced and solved for the electron-phonon scattering case. The famous Grüneisen formula of electrical resistivity is derived.

Stress is on the foundational issues like when and why Relaxation Time Approximation (RTA) is valid and why it is called the RTA. The reasons for going beyond RTA are given, and after introducing the Boltzmann equation beyond RTA it is shown how it takes quantum statistical effects into account.

Before we introduce the Boltzmann equation, some preliminary considerations are needed which are given in the next section.

2.1 Semiclassical model of electron dynamics

It is straightforward to justify the application of the classical Boltzmann equation to gas molecules as they can be treated as classical particles to a good approximation in the dilute gas limit. But it is not so straightforward to justify the application of the Boltzmann equation to electrons in metals. Because, in a metal, electrons do not behave like classical particles. The failure of the classical Drude model in this context is well known (in addressing the electronic specific heat as discussed in the previous chapter). The contributions of the Sommerfeld and Bloch, which use quantum instead of classical concepts, are surveyed in the last chapter. Electrons in metals have to be treated quantum mechanically (the Fermi-Dirac distribution of electrons addresses the specific heat problem) and according to the fundamental thesis of Bloch, electrons in an ideal lattice behave much like waves sneaking through the lattice without suffering scattering from ion cores† in contrast to the scattering picture of Drude.

Below we present some arguments that justify the use of the Boltzmann equation to study dynamics of electrons in metals. As is well known, the solution of the Schroedinger equation for electrons in a periodic lattice potential leads to the energy band picture of solids. Electrons behave as special waves called Bloch waves (due to the periodicity of the lattice potential‡). The electronic eigenenergies $\xi_n(\mathbf{k})$ and eigenfunctions $\psi_n(\mathbf{k})$ (Bloch functions) depend on the band index n and on the wavevector $\mathbf{k}$. By treating electrons as Bloch waves one can explain the thermodynamical properties of electrons in metals very well.

But the discussion of the transport properties is not so straightforward. We first write down the semiclassical equations of motion and their justification is given afterward. The effective momentum of an electron in the periodic potential is given as $\hbar\mathbf{k}$. This already takes into account the periodic potential seen by the electron. Let the sample (a piece of metal) be under the action of an external electromagnetic field. The effective momentum of the electron will evolve according to Newton's law

$$\hbar\dot{\mathbf{k}} = -e\left(\mathbf{E}(\mathbf{r},t) + \frac{1}{c}(\mathbf{v}_n(\mathbf{k}) \times \mathbf{H}(\mathbf{r},t))\right). \tag{2.1}$$

Here the electron only "sees" the external field (the internal periodic field due to ions is already taken into account in the solution of the Schroedinger equation that is contained in the effective momentum and velocity $\mathbf{v}_n(\mathbf{k})$). The band velocity $\mathbf{v}_n(\mathbf{k})$ is given by the derivative $\frac{1}{\hbar}\frac{\partial\xi_n(\mathbf{k})}{\partial\mathbf{k}}$. Thus, the position of

†This is true for an ultra pure sample. In a realistic material there are imperfections and impurity atoms etc. that act like scattering centers. And at finite temperatures thermal vibrations of atoms about equilibrium positions can also scatter electrons (we are going to study this in detail in the Bloch-Boltzmann theory).

‡If the potential is uniform in space, electron waves are the plane waves.

an electron evolves according to

$$\dot{\mathbf{r}}(t) = \mathbf{v}_n(\mathbf{k}) = \frac{1}{\hbar}\frac{\partial \xi_n(\mathbf{k})}{\partial \mathbf{k}}. \tag{2.2}$$

The above equations treat electrons as localized entities (wavepackets) which obey the Newton laws because of finite charge and mass.[§] This stands in contradiction with the picture that was given before in which electrons were treated as waves (plane waves for uniform lattice potential and Bloch waves for periodic lattice potential). In the wave picture, electron waves were treated as coherently sneaking through the lattice potential without suffering scattering. So we have changed the fundamental picture from electron waves to electron wavepackets.

Justification of this picture from the first principles is difficult. As argued in the book [1], one way to agree on this is to consider this semiclassical model and then work out its consequences and compare with the experiments–a posteriori justification.[¶] We are also going to consider this approach. So the picture that we have to keep in our minds is that of the electron wavepackets obeying the Newton laws of motion, and for the semiclassical picture the external field is treated as a classical field. One can, however, make this picture plausible by using the experimental fact. The elective momentum $\hbar\mathbf{k}$ is a good quantum number; thus, uncertainty in it must be tiny as compared to its magnitude. This uncertainty originates from the presence of collisions with imperfections and/or phonons. If we take the momentum magnitude of the order of zone momentum k_F, then $\Delta x \Delta(\hbar k) \sim \hbar, \quad \Rightarrow \quad \Delta x \sim \frac{\hbar}{\hbar \Delta k} >> \frac{1}{k_F} \sim a$, where a is the lattice constant. Thus, $\Delta x >> a$, or the size of the electron wavepacket is much bigger than the lattice constant. Thus, Figure 2.1 is what we have to keep in our minds. With this picture of Electron Wavepackets (EWs) treated semiclassically under the action of an external classical field we

[§]The position vector in the above equation has something to do with some localized entity.

[¶]The current situation is equivalent to that in statistical mechanics. There one uses the fundamental assumption of "equal a priori probabilities" to justify the microcanonical

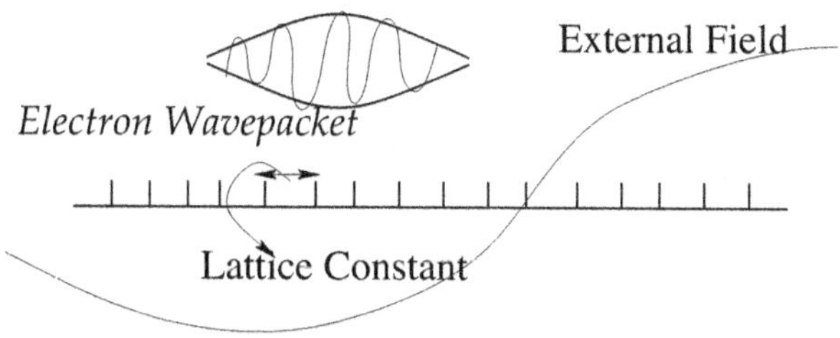

FIGURE 2.1: The semiclassical model of the electron dynamics. The lattice constant is the smallest length scale and the wavelength of the external field is the largest one. The size of the electron wavepacket lies in between.

can now apply the Boltzmann equation to this problem, and we are enabled to define the distribution function $f(\mathbf{r}, \mathbf{k}, t)$.[‖] The only issue that is remaining is to treat the scattering of these EWs. Scattering is treated phenomenologically (in the present section) by introducing a mean scattering rate that can come from the scattering of EWs by the impurity atoms or by the phonons. These cases are analyzed in detail in the subsequent sections.

To discuss Boltzmann equation we present here its reformulation by R. G. Chambers [3].

---

ensemble. In statistical mechanics too the justification comes only a posteriori–in view of the great success that statistical mechanics enjoys (see [2] for related issues).

[‖] $\mathbf{r}$ and $\mathbf{k}$ do not commute with each other, so this definition appears unjustified. But as stated above, the uncertainty in $\mathbf{k}$ is much smaller than its magnitude k_F. This leads to uncertainty in size, and we observe $\Delta x >> a$. If we agree that measurement accuracy of $\mathbf{r}$ is much bigger than the size of the electron wave packet, then we are in semi-classical limit and our definition of the distribution function is justified.

2.2 Chambers' method for Boltzmann kinetic equation (relaxation time approximation)

In the absence of an external deriving field the system is in equilibrium. The equilibrium distribution is given by the Fermi-Dirac distribution function $f(\xi_n(\mathbf{k}))$. The number of electrons in the volume element d^3k around $\mathbf{k}$ is

$$2f(\xi_n(\mathbf{k}))\frac{d^3k}{(2\pi)^3} = \frac{1}{e^{\beta(\xi_n(\mathbf{k})-\mu)}+1}\frac{d^3k}{4\pi^3}. \tag{2.3}$$

The factor of 2 is to account for spin degeneracy and $\beta = \frac{1}{k_BT}$.

When a system is under the action of an external electric field, say, for example, it no longer remains in equilibrium and deviates from the equilibrium distribution. Let the nonequilibrium distribution be denoted by $g(\mathbf{r}, \mathbf{k}, t)$.** This acts on a semiclassical phase space whose volume element is denoted as $d^3r(t)d^3k(t)$ such that the number of electrons in this volume element at time t is given by

$$dN = 2g(\mathbf{r}, \mathbf{k}, t)\frac{d^3r(t)d^3k(t)}{(2\pi)^3}. \tag{2.4}$$

The time evolution of $\mathbf{r}(t)$ and $\mathbf{k}(t)$ is given by the semiclassical equations of motion (Equations (2.1) and (2.2)). We are going to compute this number based on the semiclassical trajectory of a phase point in the phase space. By doing that we will be able to express the unknown function $g(\mathbf{r}, \mathbf{k}, t)$ in terms of the known functions. The phase space that we consider is the one electron phase space (μ-space) of 6 dimensions (three for coordinates and remaining three for the momenta††). Thus, for each electron there is a phase point and for N electrons there is a swarm of phase points (in Figure 2.2) a single trajectory, for simplicity, is depicted). Now the number of electrons dN in the

**For simplicity we suppress the band index n. There is no loss of generality. If needed for explicit purpose we will reinsert the subscript n.

††It is different from the phase space Γ which is $6N$ dimensional and in this Γ−space the state of the whole system (composing of N particles) is represented by a point.

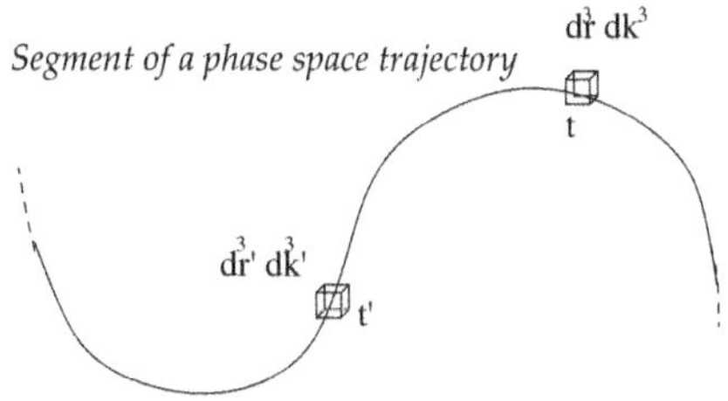

FIGURE 2.2: A segment of the trajectory of the phase point in the $\mu-$phase space (phase space of a single electron of 6 dimensions). The phase trajectory is governed by the semiclassical equations of motion.

nonequilibrium state present at time t in the volume element $d^3r(t)d^3k(t)$ can be computed as follows. Let us consider a phase trajectory that passes through the phase point $(\mathbf{r}(t),\ \mathbf{k}(t))$ at time t and $d^3r(t)d^3k(t)$ is the differential volume surrounding it. Assume that at some past time t' $(t' < t)$ it passed through a point $(\mathbf{r}(t'),\ \mathbf{k}(t'))$ and $(d^3r(t')d^3k(t'))$ is the differential volume element that surrounds it (refer to Figure 2.2). This trajectory is deterministic and is given by the semiclassical equations of motion. If $\tau(\mathbf{r}(t'), \mathbf{k}(t'))$ is the phase point dependent relaxation time, then the number of electrons scattered off from the volume element $d^3r(t')d^3k(t')$ in the time interval dt' around t' is

$$\frac{dt'}{\tau(\mathbf{r}(t'), \mathbf{k}(t'))} g_{eq}(\mathbf{r}(t'), \mathbf{k}(t'))d^3r(t')d^3k(t'). \tag{2.5}$$

This is based on the "method of tube-integrals" as devised by Chambers [3] (refer for more details [4, 5, 6, 7]). *In this way of counting, one assumes that the system is in equilibrium.* The philosophy behind this method is to investigate carefully the mechanism by which the system initially in equilibrium state goes to the nonequilibrium state under external perturbation. The above is the number of electrons that are scattered out of the equilibrium distribution $g_{eq}(\mathbf{r}(t'), \mathbf{k}(t'))$ at $(\mathbf{r}(t'), \mathbf{k}(t'))$ in the time interval from t' to $t' + dt'$, and goes into a nonequilibrium distribution $g(\mathbf{r}(t'), \mathbf{k}(t'))$ which is not known to us. Now, if these electron, afterward, reach (via semiclassical dynamics) to volume element $d^3r(t)d^3k(t)$ *without suffering any collision*, then, by summing over all t' less than t it is possible to calculate dN, provided the survival probability is known[‡‡]

$$dN = \int_{-\infty}^{t} \frac{dt'}{\tau(\mathbf{r}', \mathbf{k}')} g_{eq}(\mathbf{r}', \mathbf{k}', t')\mathcal{P}(\mathbf{r}, \mathbf{k}, t; t') \frac{d^3r'd^3k'}{4\pi^3}. \tag{2.6}$$

Here we have used the notation $\mathbf{r}' = \mathbf{r}(t')$ and $\mathbf{k}' = \mathbf{k}(t')$. The survival probability pertaining to volume element d^3rd^3k is denoted by $\mathcal{P}(\mathbf{r}, \mathbf{k}, t; t')$. Survival probability is the probability that an electron goes from t' to t without suffering any collision. The average time interval between the collisions $\tau(\mathbf{r}', \mathbf{k}')$ is taken to be phase point dependent. Then by the Liouville theorem one must have $d^3r'd^3k' = d^3rd^3k$ (i.e., semiclassical dynamics preserves the phase space volume). Thus, from Equations (2.4) and (2.6) we obtain a closed expression for the nonequilibrium distribution function

$$g(\mathbf{r}, \mathbf{k}, t) = \int_{-\infty}^{t} \frac{dt'}{\tau(\mathbf{r}', \mathbf{k}')} g_{eq}(\mathbf{r}', \mathbf{k}', t')\mathcal{P}(\mathbf{r}, \mathbf{k}, t; t'). \tag{2.7}$$

It is possible to numerically/analytically solve this integral equation once average time between collisions and the survival probability are given. From the nonequilibrium distribution function mean value of any one particle observable (in the nonequilibrium state) can be computed

$$\langle O(t)\rangle = \int d^3rd^3k\ g(\mathbf{r}, \mathbf{k}, t)O(\mathbf{r}, \mathbf{k}), \tag{2.8}$$

For example, mean kinetic energy of an electron, etc. Equation (2.7) is the central result and is called Chambers' kinetic integral in Relaxation Time Approximation (RTA). In the next section we will derive this result from the full Boltzmann equation, where physical assumptions for RTA will be more clear.

To proceed further it will be convenient to shorten the notation a

[‡‡]The assumption of ergodicity is assumed in which over a very long time interval the phase trajectory traverses the whole phase space. The limits of the integration from $-\infty$ to t ensure this.

bit. We will use the following notation: $\tau(\mathbf{r}',\mathbf{k}') = \tau(\mathbf{r}(t'),\mathbf{k}(t')) \to \tau(t')$, $g_{eq}(\mathbf{r}',\mathbf{k}') \to g_{eq}(t')$, $\mathcal{P}(\mathbf{r},\mathbf{k},t;t') \to \mathcal{P}(t;t')$. Under this, the above equation reduces to

$$g(t) = \int_{-\infty}^{t} dt' g_{eq}(t') \frac{\mathcal{P}(t;t')}{\tau(t')}. \tag{2.9}$$

To construct a differential equation for the survival probability consider again the time interval from t' to t ($t' < t$). Now, the chance that an electron emerging at the instant t' will not have any collision in the subsequent infinitesimally small time interval dt' is $1 - \frac{dt'}{\tau(t')}$. Here $\frac{dt'}{\tau(t')}$ is the probability that the electron will suffer collision. Thus

$$(1 - dt'/\tau(t'))\mathcal{P}(t' + dt', t) = \mathcal{P}(t', t), \tag{2.10}$$

i.e., survival probability from t' to $t'+dt'$ multiplied by the survival probability from $t' + dt'$ to t is equal to the survival probability from t' to t. From the above equation it is easy to prove that

$$\lim_{dt' \to 0} \frac{\mathcal{P}(t' + dt', t) - \mathcal{P}(t', t)}{dt'} = \frac{\partial \mathcal{P}(t', t)}{\partial t'} = \frac{\mathcal{P}(t', t)}{\tau(t')}. \tag{2.11}$$

Substituting $\frac{\mathcal{P}(t',t)}{\tau(t')} = \frac{\partial \mathcal{P}(t',t)}{\partial t'}$ into Equation (2.9) and then integrating by parts leads to

$$g(t) = g_{eq}(t) - \int_{-\infty}^{t} dt' \mathcal{P}(t', t) \frac{dg_{eq}(t')}{dt'}, \tag{2.12}$$

where the following conditions are used: $\mathcal{P}(t,t) = 1$, $\mathcal{P}(-\infty, t) = 0$ (full survival when the time interval is zero and no survival when the time interval is infinity). The equilibrium distribution function depends on time through $\mathbf{r}$ and $\mathbf{k}$ thus

$$\frac{dg_{eq}(t)}{dt} = \frac{\partial g_{eq}}{\partial \xi} \frac{\partial \xi}{\partial \mathbf{k}} \cdot \frac{d\mathbf{k}}{dt} + \frac{\partial g_{eq}}{\partial T} \frac{\partial T}{\partial \mathbf{r}} \cdot \frac{d\mathbf{r}}{dt} + \frac{\partial g_{eq}}{\partial \mu} \frac{\partial \mu}{\partial \mathbf{r}} \cdot \frac{d\mathbf{r}}{dt}. \tag{2.13}$$

Using semiclassical equations of motion (2.1) and (2.2) it is easy to prove that Equation (2.12) can be reduced to

$$\boxed{g(t) = g_{eq}(t) + \int_{-\infty}^{t} dt' \mathcal{P}(t', t) \left[(-\frac{\partial g_{eq}}{\partial \xi}) \mathbf{v}.(-e\mathbf{E} - \nabla \mu - (\frac{\xi - \mu}{T}) \nabla T) \right]}. \tag{2.14}$$

Thus, nonequilibrium distribution function can be calculated if the survival probability and external driving fields are given. From nonequilibrium distribution function mean value of any phase function (observable) can be computed.

2.2.1 Obtaining Drude's formula for conductivity from the Boltzmann equation

In this section we will simplify Equation (2.14) for the case of electrical conduction in metals.

2.2.1.1 DC transport

Consider a metallic sample under the bias of a DC electric field. Consider also that the sample is kept at a uniform temperature and uniform chemical potential (i.e., $\nabla\mu = \nabla T = 0$). We have used the assumption of mean collision time. If the density and temperature of the electron gas do not change with time, then the mean collision time will not depend on time. Thus, in a steady state situation τ is independent of time. It can, however, depend on the energy. Since in metals conduction electrons are from a tiny zone of size $\sim k_BT$ around the Fermi surface of energy scale E_F with $k_BT << E_F$, the mean collision time can also be taken as energy independent. Under these physical conditions, Equation (2.11) gives the survival probability as

$$\mathcal{P}(t', t) = e^{-\frac{t-t'}{\tau}}. \tag{2.15}$$

Thus, chances of survival exponentially fall. And Equation (2.14) reduces to

$$g = g_{eq} - \tau e\mathbf{v}.\mathbf{E}(-\frac{\partial g_{eq}}{\partial \xi}). \tag{2.16}$$

Here g is only $\mathbf{k}$ dependent (electric field is uniform, no $\mathbf{r}$ dependence). Under the action of the electric field there is a steady current flow in the sample the current density of which is

$$\mathbf{J} = -2e \int \frac{d^3k}{(2\pi)^3} \mathbf{v}(\mathbf{k}) g(\mathbf{k}). \tag{2.17}$$

Inserting the equation for g (Equation 2.16) into the above equation leads to two terms. The first will be the value of current in equilibrium and it vanishes due to obvious reasons and the second gives the conductivity tensor

$$\mathbf{J} = \sigma \mathbf{E}, \qquad \sigma = e^2\tau \int \frac{d^3k}{4\pi^3}\left(-\frac{\partial g_{eq}}{\partial \xi}\right)\mathbf{v}(\mathbf{k})\mathbf{v}(\mathbf{k}). \tag{2.18}$$

For simplicity, consider that an electric field is applied in the x-direction and the current is also measured in the same direction. Then the conductivity is given by

$$\sigma_{xx} = e^2\tau \int \frac{d^3k}{4\pi^3}\left(-\frac{\partial g_{eq}}{\partial \xi}\right)v_x^2(\mathbf{k}). \tag{2.19}$$

At low temperatures $k_BT << E_F$ the derivative $-\frac{\partial g_{eq}}{\partial \xi}$ is effectively a delta function ($\delta(\xi - E_F)$) around the Fermi surface. And in metals electrons can be treated as free, thus $\xi(\mathbf{k}) = \frac{\hbar^2 k^2}{2m}$ and thus $\mathbf{v} = \frac{1}{\hbar}\frac{\partial \xi}{\partial \mathbf{k}} = \frac{\hbar \mathbf{k}}{m}$. Due to isotropy $v_x^2 = \frac{1}{3}v^2$. Collecting this information and converting the integral from $k-$space to $\xi-$space using the transformation $k = \frac{\sqrt{2m\xi}}{\hbar}$ and using $k_F^3 = 3\pi^2 n$ conductivity can be shown to obey

$$\sigma_{xx} = \frac{ne^2\tau}{m}. \tag{2.20}$$

This is nothing but the Drude formula for conductivity obtained previously using Newton's law of motion with a dissipation term. Thus, we have $\tau = \tau_{Drude}$. The average rate at which the momentum is degraded, i.e., $1/\tau_{Drude}$ in the Drude model is the same as the rate at which the survival probability is degraded in the Boltzmann equation. Most importantly, this is also the time scale in which the system initially in the nonequilibrium state relaxes to the equilibrium state.

To understand the last statement let us consider that the electric field is applied adiabatically (the strength of the field is very slowly increased up to its maximum value) and then it is switched off at $t = 0$. In this way the system was slowly brought to a nonequilibrium state in which the current was flowing, then at $t = 0$ the field is switched off. We are interested in how the system will relax to equilibrium after $t = 0$, i.e., what will be the time scale in which the system will regain its equilibrium (condition of no current flow).

Let the field be given by $E = E_0^x e^{\epsilon t}\Theta(-t)$ for $-\infty < t < 0$. Here $\epsilon \to 0$ (to ensure slow rise in the field strength). In the present case, we will have (instead of Equation (2.16))

$$g = g_{eq} - \tau e^{-t/\tau} e\mathbf{v}.\mathbf{E}(-\frac{\partial g_{eq}}{\partial \xi}). \tag{2.21}$$

The current density J_x simplifies to

$$J_x = 2e^2\tau e^{-t/\tau} E_0^x \int \frac{d^3k}{(2\pi)^3} v_x^2(\mathbf{k})\delta(\xi - E_f) = \frac{ne^2\tau}{m} E_0^x e^{-t/\tau}. \tag{2.22}$$

Here we again used the free electron model and spherical Fermi surface ($\xi = \frac{\hbar^2 k^2}{2m}$). The $k-$space integral is converted to energy integral as before, and the algebra is the same as done for the conductivity (refer to problem number (2) in the last chapter).

We notice that, after the field is switched off at $t = 0$, the current density exponentially reduces to zero in a time scale of τ. Thus, the relaxation time of the system is τ. *From this a very important physical conclusion is reached: the mean collision time τ is equal to the relaxation time scale of the system.* Thus, within a few collisions between the particles, the system reaches equilibrium.

2.2.1.2 AC transport

Consider that the applied field has now frequency dependence $E = E_0^x \cos(\omega t)$. On the similar lines and assumptions made in the previous section, it is easy to show that nonequilibrium distribution function (Equation 2.16) takes the form

$$g(t) = g_{eq} - \tau e v_x E_0^x (-\frac{\partial g_{eq}}{\partial \xi})\left(\frac{1}{1+\omega^2\tau^2}\cos(\omega t) + \frac{\tau\omega}{1+\omega^2\tau^2}\sin(\omega t)\right). \tag{2.23}$$

The calculation of current density is also along similar lines and AC conductivity takes the form

$$\sigma_{xx}(\omega) = \frac{ne^2\tau}{m}\left(\frac{1}{1+\omega^2\tau^2}\cos(\omega t) + \frac{\tau\omega}{1+\omega^2\tau^2}\sin(\omega t)\right). \tag{2.24}$$

As we have seen in Section 1.3 of the last chapter, it has two components. The in-phase component (the dissipative component) and the out-of-phase component (the reactive response). The reason is explained in Section 1.3. Also for $\omega \to 0$ the AC conductivity reduces to the DC conductivity, as it should (refer to problem number (3)).

2.3 Beyond the relaxation time approximation (full Boltzmann equation)

In the last section we used the assumption of the existence of an average time interval between the collisions $\tau(\mathbf{r}, \mathbf{k})$, i.e., the Assumption of the Mean Collision Time (AMCT). The validity of this assumption is based on two fundamental ingredients: (1) the law of large numbers, and (2) independence of the collision processes from the quantum statistics. This leads to an integral equation (2.7) for $g(\mathbf{r}, \mathbf{k}, t)$. In the quantum limit, the probability of scattering of an electron from k state into k' state depends heavily on the occupation of the k' state (the Pauli exclusion principle). In RTA we neglected this very important fact. And it was justified for a semiclassical gas of electrons when mean occupation of any $k-$state is much less than one.

More accurate treatment must consider the Pauli exclusion principle. The scattering rate then takes the form

$$\boxed{\frac{1}{\tau(\mathbf{k})} = \int \frac{d^3k'}{(2\pi)^3} W_{\mathbf{k},\mathbf{k}'}(1 - g(\mathbf{k}')).} \tag{2.25}$$

Here $\frac{1}{\tau(\mathbf{k})}$ is the scattering rate from $k-$state, $W_{\mathbf{k},\mathbf{k}'}$ represents the scattering rate from $k-$state to $k'-$state per unit $k-$space volume, and the factor $(1 - g(\mathbf{k}'))$ represents the probability that $k'-$state is empty. If it is not empty, electron won't scatter into it (the Pauli exclusion principle). The scattering rate is obtained by integrating over all possible final $k'-$states. If $g(\mathbf{k}) << 1$, the scattering rate becomes distribution independent and we get back the RTA form.

Next, we want to compute the rate of change of the nonequilibrium distribution function due to collisions. By equating it to a term that originates due the application of an external field (the drift term) one obtains the Boltzmann equation. For the computation of the collision term let us compute the number of electrons that are scattered out of $k-$state in a time interval Δt. Let us denote this number by $\Delta t(\frac{dg(\mathbf{k})}{dt})_{out}$. Clearly

$$\left(\frac{dg(\mathbf{k})}{dt}\right)_{out}\Delta t = -\frac{g(\mathbf{k})}{\tau(\mathbf{k})}\Delta t = -g(\mathbf{k})\Delta t\int\frac{dk'^3}{(2\pi)^3}W_{\mathbf{k},\mathbf{k}'}(1-g(\mathbf{k}')), \quad (2.26)$$

or

$$\left(\frac{dg(\mathbf{k})}{dt}\right)_{out} = -g(\mathbf{k})\int\frac{d^3k'}{(2\pi)^3}W_{\mathbf{k},\mathbf{k}'}(1-g(\mathbf{k}')). \quad (2.27)$$

At the same time, $k-$state is also gaining electrons (the electrons which scatter from any k' to k). This number is denoted by $(\frac{dg(\mathbf{k})}{dt})_{in}\Delta t$ and is computed as follows. Let the scattering rate from k' to k per unit $k-$space volume be denoted as $W_{\mathbf{k}',\mathbf{k}}$. This transition is possible if $k'-$state is occupied and $k-$state is empty. By multiplying these probabilities and then summing over all possible k' one obtains

$$\left(\frac{dg(\mathbf{k})}{dt}\right)_{in}\Delta t = \Delta t\int\frac{d^3k'}{(2\pi)^3}W_{\mathbf{k}',\mathbf{k}}g(\mathbf{k}')(1-g(\mathbf{k})), \quad (2.28)$$

or

$$\left(\frac{dg(\mathbf{k})}{dt}\right)_{in} = \int\frac{d^3k'}{(2\pi)^3}W_{\mathbf{k}',\mathbf{k}}g(\mathbf{k}')(1-g(\mathbf{k})). \quad (2.29)$$

Thus, the rate of change of nonequilibrium distribution function due to collisions is given by

$$\left(\frac{dg(\mathbf{k})}{dt}\right)_{collisions} = \left(\frac{dg(\mathbf{k})}{dt}\right)_{in} + \left(\frac{dg(\mathbf{k})}{dt}\right)_{out}. \quad (2.30)$$

In the case of an applied electric field $g(k)$ will also evolve with time. To obtain the Boltzmann equation first consider the collisionless gas. Due to the Liouville theorem $\mu-$space volume is conserved

$$\frac{dg}{dt} = 0, \text{or}, \quad \frac{\partial g}{\partial t} + \frac{\partial g}{\partial \mathbf{r}}.\frac{d\mathbf{r}}{dt} + \frac{\partial g}{\partial \mathbf{k}}.\frac{d\mathbf{k}}{dt} = 0. \quad (2.31)$$

If collisions are present then $\frac{dg}{dt} = (\frac{dg}{dt})_{collisions}$ and the Boltzmann equation can be written in its popular form

$$\boxed{\frac{\partial g}{\partial t} + \frac{\partial g}{\partial \mathbf{r}}.\mathbf{v} + \frac{\partial g}{\partial \mathbf{k}}.\frac{\mathbf{F}}{\hbar} = \left(\frac{dg(\mathbf{k})}{dt}\right)_{collisions} = -\int \frac{d^3k'}{(2\pi)^3} W_{\mathbf{k},\mathbf{k}'}(g(\mathbf{k}) - g(\mathbf{k}')).} \tag{2.32}$$

Here we have used $W_{\mathbf{k},\mathbf{k}'} = W_{\mathbf{k}',\mathbf{k}}$ the law of microscopic reversibility.[§§] The terms arising due to external field $\mathbf{F}$[¶¶] and spatial inhomogeneity are called the drift terms and the integral containing collision terms is called the collision integral. The Boltzmann equation is an integro-differential equation and its direct analytical solutions in full generality are not possible. It is generally solved numerically, for example, in the case of nonequilibrium transport in semiconductors there is a large amount of literature on it (refer to [8] and references therein). But for phonon and impurity scattering cases, it is possible to obtain the collision term analytically and these are considered in detail in the next sections. We will end this section by reconsidering the RTA and considering a special and important case of elastic scattering in isotropic medium.

2.3.1 Physical assumptions for RTA

The beyond RTA Boltzmann equation (2.32) reduces to its familiar RTA form if (1) there exists a universal scattering rate which is independent of the nonequilibrium distribution function (only governed by the Local Thermodynamic Equilibrium (LTE) form of the distribution function), (2) deviations from LTE form are weak, (3) $|g(\mathbf{k})| << 1$ so that quantum statistical effects can be neglected (dilute gas), and (4) elastic scattering or weak inelastic scattering is the cause of momentum randomization.

[§§]It can be explicitly seen for the impurity scattering case $W_{\mathbf{k},\mathbf{k}'} = \frac{2\pi}{\hbar} n_i \delta(\xi(\mathbf{k}) - \xi(\mathbf{k}'))|\langle\mathbf{k}|U|\mathbf{k}'\rangle|^2$, delta function is even under the change of k to k' and the matrix element is symmetric under this change. The law is generally valid [1]. Also see the Bloch-Boltzmann theory in the next section.

[¶¶]Important case generally studied is $\mathbf{F} = -e(\mathbf{E} + (1/c)(\mathbf{v} \times \mathbf{H}))$.

Under these conditions Equations (2.27) and (2.29) can be written as

$$\left(\frac{dg(\mathbf{k})}{dt}\right)_{out} = -\frac{g(\mathbf{k})}{\tau_{RTA}(\mathbf{k})}, \quad \left(\frac{dg(\mathbf{k})}{dt}\right)_{in} = \frac{g_{eq}(\mathbf{k})}{\tau_{RTA}(\mathbf{k})}. \tag{2.33}$$

Here $\tau_{RTA}(\mathbf{k}) = \int \frac{d^3k'}{(2\pi)^3} W_{\mathbf{k},\mathbf{k}'}$. To obtain the above equations replace $1 - g(\mathbf{k}')$ and $1 - g(\mathbf{k})$ with 1 in Equations (2.27) and (2.29)[mean occupation much smaller than one]. For the second equation use $g(\mathbf{k}') \simeq g_{eq}(\mathbf{k}')$ in the integrand for scattering rate, and then use the condition for elastic scattering $k = k'$.*** On performing these steps, the Boltzmann equation takes the familiar form

$$\boxed{\frac{\partial g}{\partial t} + \frac{\partial g}{\partial \mathbf{r}}.\mathbf{v} + \frac{\partial g}{\partial \mathbf{k}}.\frac{\mathbf{F}}{\hbar} = \left(\frac{dg(\mathbf{k})}{dt}\right)_{collisions} = -\frac{g(\mathbf{k}) - g_{eq}(\mathbf{k})}{\tau_{RTA}(\mathbf{k})}.} \tag{2.34}$$

If external fields are switched off the system relaxes to equilibrium in a time scale of τ_{RTA} (this can be easily verified by solving collision term which constitutes first order differential equation [last two terms in the above equation]).

The above equation reduces to Equation (2.21) obtained by Chamber's method when electric field is applied (refer to practice exercise (4) in Chapter 8).

2.3.2 Elastic scattering in an isotropic medium

The case of elastic scattering in an isotropic medium has a special feature. Let us perform the following iteration. If nonequilibrium distribution function from RTA in an isotropic medium is used in the Collision Integral (CI) of the full (beyond RTA) Boltzmann equation, then CI of the full Boltzmann equation can be simplified using properties of elastic scattering and it can be

***From the context it appears that the condition of elastic scattering is too restrictive and RTA can be justified only of elastic impurity scattering. However, one should note that even in the case of electron phonon scattering, the change in the energy of the scattered electron is very small as compared to the kinetic energy of the electron itself. This is due to the fact that phonon energies involved are very small ($\sim 10^{-3}\ eV$) as compared to electronic energies ($\sim 1\ eV$). Hence the scattering can be considered approximately elastic.

cast in the form of RTA CI with a special scattering rate

$$\frac{1}{\tau_{tr}(\mathbf{k})} = \int \frac{d^3k'}{(2\pi)^3} W_{\mathbf{k},\mathbf{k}'}(1 - \cos(\theta)). \tag{2.35}$$

Here θ is the angle between $\mathbf{k}$ and $\mathbf{k}'$ (the direction of $\mathbf{k}$ can be taken along the z-axis for evaluating the integral). To prove this, substitute TRA g from Equation (2.16) into the collision integral of the full Boltzmann Equation (2.32). The RTA g from Equation (2.16) for an isotropic medium ($\mathbf{v}(\mathbf{k}) = \hbar\mathbf{k}/m$) can be written as

$$g(\mathbf{k}) = g_{eq}(|\mathbf{k}|) + \mathbf{a}(|\mathbf{k}|).\mathbf{k}, \tag{2.36}$$

where $\mathbf{a}(|\mathbf{k}|)$ is a vector function of $|\mathbf{k}|$ which is equal to $-\frac{\tau e\hbar}{m}\mathbf{E}(\partial g_{eq}(\xi)/\partial\xi)$ and $\xi = \frac{\hbar^2 k^2}{2m}$ is a scalar function of k. Substitute this into the collision integral of the full Boltzmann equation (2.32)

$$\int \frac{d^3k'}{(2\pi)^3} W_{\mathbf{k},\mathbf{k}'}(\mathbf{a}(|\mathbf{k}|).\mathbf{k} - \mathbf{a}(|\mathbf{k}'|).\mathbf{k}' + g_{eq}(|\mathbf{k}|) - g_{eq}(|\mathbf{k}'|)), \tag{2.37}$$

which reduces to

$$\mathbf{a}(k).\left(\int \frac{d^3k'}{(2\pi)^3} W_{\mathbf{k},\mathbf{k}'}\ (\mathbf{k} - \mathbf{k}')\right), \tag{2.38}$$

for the case of elastic scattering (where magnitude of $\mathbf{k}$ is conserved, i.e., $k = k'$) and g_{eq} depends only on the magnitude of $\mathbf{k}$. Suppose that the above equation can be written in the form

$$\frac{1}{\tau_{tr}(\mathbf{k})}(g(\mathbf{k}) - g_{eq}(k)). \tag{2.39}$$

Our aim is to find $\tau_{tr}(\mathbf{k})$ if the above two equations are equal. By equating the above two equations and substituting for $g(\mathbf{k})$ from Equation (2.36) into Equation (2.39), we obtain

$$\frac{\mathbf{k}}{\tau_{tr}(\mathbf{k})} = \int \frac{d^3k'}{(2\pi)^3} W_{\mathbf{k},\mathbf{k}'}\ (\mathbf{k} - \mathbf{k}'). \tag{2.40}$$

Take the dot product with $\mathbf{k}$ on both sides and by noting that $k = k'$ (magnitudes are equal), we obtain

$$\frac{1}{\tau_{tr}(\mathbf{k})} = \int \frac{d^3k'}{(2\pi)^3} W_{\mathbf{k},\mathbf{k}'}\ (1 - \cos\theta), \tag{2.41}$$

where the angle between $\mathbf{k}$ and $\mathbf{k}'$ is θ (practice exercise 5).

Thus, the CI of the full Boltzmann equation can be written as the RTA form $\frac{g(k)-g_{eq}(k)}{\tau_{tr}(\mathbf{k})}$ with a special scattering rate given by Equation (2.35). This scattering rate brings about an important point that the forward scattering events (when $\theta = 0$) are not operative in the momentum degradation of a quasiparticle and the large angle scatterings are important (back scattering with the maximum effect).

In the general case of an inelastic scattering in an anisotropic medium it is not possible to write a compact expression like the above along these lines but General conclusion mentioned above appears to hold well [4].

2.4 The Bloch-Boltzmann transport theory

Thus far we have presented a general introduction to the Boltzmann equation, and impurity scattering case was considered. This section is devoted to a very important achievement by Felix Bloch in which he resolved the long-standing problems of Boltzmann-Drude-Lorentz theory. As we learned, in the first chapter, the Drude theory and its subsequent formulation in terms of the Boltzmann equation by Lorentz was not able to address the temperature dependence of electrical and thermal conductivities. The crux of the matter was in the proper quantum mechanical treatment of electron-lattice interaction which was achieved by Bloch. In this section we are going to present the Bloch-Boltzmann theory and we will observe how the theory is able to explain the experimentally observed temperature evolution of DC resistivity:

$$\rho(T) \sim \begin{cases} T & : T >> \Theta_D \\ T^5 & : T << \Theta_D. \end{cases} \tag{2.42}$$

Our plan for this section is the following. First we will collect all the assumptions in one place at the beginning. Then the Boltzmann equation will be written with general collision integral. The collision integral for the electron-phonon interaction will be written using first order perturbation theory, i.e., using Fermi's golden rule. Next, the collision integral will be simplified using the properties of the Dirac delta functions, and it will be written in a symmetric form. Then the collision integral will be computed for the special case of electrical conductivity. Explicit calculation will be done with an "iterative" procedure. Finally, the collision integral will be simplified for the special cases of low and high temperatures, and the resistivity will be computed. Let us first start by collecting all the assumptions involved:

ASSUMPTIONS:

1. About electrons: Electrons are represented as plane waves $\phi(\mathbf{r}) = \frac{1}{\sqrt{V}}e^{-i\mathbf{k}.\mathbf{r}}$ with electronic energy $E_{\mathbf{k}} = \frac{\hbar^2 k^2}{2m}$. We neglect electron-electron interactions and thus assume them free. They constitute a spherical Fermi surface with radius k_F and their thermal equilibrium distribution is given by

$$f_{\mathbf{k}} = \frac{1}{e^{\beta(E_{\mathbf{k}}-\mu)}+1}, \tag{2.43}$$

where $\beta = \frac{1}{k_B T}$ and μ is the Fermi energy.

2. About phonons: We consider acoustic longitudinal phonons. Assume that there is one atom per unit cell and thus there are no optical phonons. The phonon dispersion is represented by $\omega_{\mathbf{q}} = s_l|q|$. Here s_l is the sound speed and $\mathbf{q}$ is the phonon wave-vector. When the system is in equilibrium (i.e., there is no external driving field) the distribution of phonons is given by Bose-Einstein distribution:

$$n_{\mathbf{q}} = \frac{1}{e^{\beta\hbar\omega_{\mathbf{q}}}-1}. \tag{2.44}$$

3. About electron-phonon interaction: We consider first order electron-phonon interactions given by Figure 2.3. And further assume that the

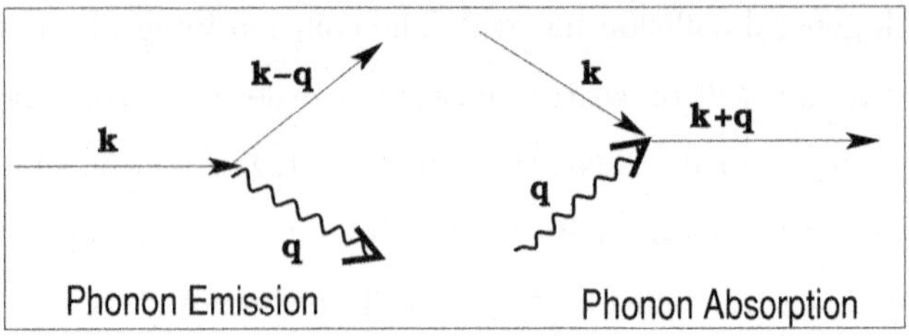

FIGURE 2.3: The phonon emission and absorption processes.

scattering preserves the electron spin. The electron-phonon interaction is given by the following Fröhlich Hamiltonian:

$$H_{el-ph} = \sum_{\mathbf{k},\mathbf{q},\sigma} M_{\mathbf{q}}(a_{\mathbf{q}} + a^{\dagger}_{-\mathbf{q}})c^{\dagger}_{\mathbf{k}+\mathbf{q},\sigma}c_{\mathbf{k},\sigma}, \quad M_{\mathbf{q}} = \frac{1}{\sqrt{2M_{ion}N\omega_{\mathbf{q}}}}qC_q. \tag{2.45}$$

Here C_q will be taken as weakly depending on q which is a reasonable assumption when the screening is strong [10].

4. *The Bloch assumption: When electric field is applied to a metallic sample, electrons absorb energy from the applied electric field and the electronic distribution does not remain the Fermi-Dirac distribution. It becomes some nonequilibrium distribution* g_k *which is our object of study. Due to electron-phonon scattering, energy absorbed by the electrons from the field transfers to the phonon bath. Thus, ideally, phonon distribution should change from the equilibrium Bose-Einstein distribution to some nonequilibrium phonon distribution. It is assumed that the internal relaxation time of the phonon bath is very short as compared to electron-phonon relaxation time. Thus, phonons very quickly redistribute their energy and reach to thermal equilibrium Bose-Einstein distribution, as given by Equation (2.44). This is called the Bloch assumption.*

Keeping these assumptions in mind we now develop the Bloch-Boltzmann theory. The Boltzmann equation with $g_{\mathbf{k}}$ as the nonequilibrium distribution function has the form:

$$\begin{aligned}\mathbf{v}(\mathbf{k}).\frac{\partial g_{\mathbf{k}}}{\partial \mathbf{r}} - \frac{\mathscr{E}}{\hbar}.\frac{\partial g_{\mathbf{k}}}{\partial \mathbf{k}} &= \left(\frac{dg_{\mathbf{k}}}{dt}\right)_{coll} = \int \frac{d^3k'}{(2\pi)^3}(gain - loss) \\ &= \int \frac{d^3k'}{(2\pi)^3}(W(\mathbf{k}' \to \mathbf{k}) - W(\mathbf{k} - \mathbf{k}')). \quad (2.46)\end{aligned}$$

Here $\mathscr{E}$ is the applied DC electric field (although it is not written in boldface, it is a vector). We substitute $\frac{\partial g_{\mathbf{k}}}{\partial t} = 0$ as we deal with a steady state situation. The $W(\mathbf{k}' \to \mathbf{k})$ measures the transition rate of an electron from a state represented by $|\mathbf{k}'\rangle$ to a state $|\mathbf{k}\rangle$. It is computed using the Fermi Golden rule:

$$W(\mathbf{k}' \to \mathbf{k}) = \frac{2\pi}{\hbar}|\langle \mathbf{k}|H_{el-ph}|\mathbf{k}'\rangle|^2 \delta(E_{\mathbf{k}} - E_{\mathbf{k}'}). \quad (2.47)$$

For the case of electron-phonon scattering we need to specify the initial state $|\mathbf{k}'\rangle$ and the final state $|\mathbf{k}\rangle$ by respecting the momentum and the energy conservation principles. Let us say that an electron scatters from a state with momentum $\mathbf{k}' = \mathbf{k} + \mathbf{q}$ to a state with momentum $\mathbf{k}$. This process can happen either by an emission of a phonon or by an absorption of a phonon. Let us set the notation in the following way:

$$\text{An electron going into state } \mathbf{k} \rightsquigarrow \begin{cases} W(\mathbf{k} + \mathbf{q} \to \mathbf{k}; \mathbf{q}) \ [Emission] \\ W(-\mathbf{q}; \mathbf{k} + \mathbf{q} \to \mathbf{k}) \ [Absorption] \end{cases}. \quad (2.48)$$

Notation means that when a phonon is emitted it is written on the right-hand side in W and when it is absorbed, it is written on the left-hand side in W. Similarly, we can write for an electron going out of state represented by $\mathbf{k}$:

$$\text{An electron going out of state } \mathbf{k} \rightsquigarrow \begin{cases} W(\mathbf{k} \to \mathbf{k} + \mathbf{q}; -\mathbf{q}) \ [Emission] \\ W(\mathbf{q}; \mathbf{k} \to \mathbf{k} + \mathbf{q}) \ [Absorption] \end{cases}. \quad (2.49)$$

With this notation the collision term of the Boltzmann equation can be written as

$$\begin{aligned}\left(\frac{dg_{\mathbf{k}}}{dt}\right)_{coll} &= \int \frac{d^3q}{(2\pi)^3}\{W(\mathbf{k} + \mathbf{q} \to \mathbf{k}; \mathbf{q}) + W(-\mathbf{q}; \mathbf{k} + \mathbf{q} \to \mathbf{k}) \\ &- W(\mathbf{k} \to \mathbf{k} + \mathbf{q}; -\mathbf{q}) - W(\mathbf{q}; \mathbf{k} \to \mathbf{k} + \mathbf{q})\}. \quad (2.50)\end{aligned}$$

We are going to compute the transition rates using the Fermi Golden rule. Below a detailed calculation for one transition rate is given, and for the others it can be inferred just by inspection. Let us consider $W(\mathbf{k}+\mathbf{q} \to \mathbf{k};\mathbf{q})$, and write the initial and final states in occupation number representation in the following way. $|initial\rangle = |N_{\mathbf{k+q}}, N_{\mathbf{k}}; n_{\mathbf{q}}\rangle$ and $|final\rangle = |N_{\mathbf{k+q}} - 1, N_{\mathbf{k}} + 1; n_{\mathbf{q}} + 1\rangle$, where $N_{\mathbf{k}}$ represents the number of electrons in state $|\mathbf{k}\rangle$ and $n_{\mathbf{q}}$ represents the number of phonons in a state represented by wave-vector $\mathbf{q}$. The final state represents that one electron is lost from state $|\mathbf{k}+\mathbf{q}\rangle$ and gained in state $|\mathbf{k}\rangle$, with the emission of a phonon. Consider the matrix element (refer to Equations (2.45) and (2.47))

$$\sum_{\mathbf{k}',\mathbf{q}'} M_{\mathbf{q}'} \langle N_{\mathbf{k+q}}-1, N_{\mathbf{k}}+1; n_{\mathbf{q}}+1|(a_{\mathbf{q}'} + a^{\dagger}_{-\mathbf{q}'})c^{\dagger}_{\mathbf{k}'+\mathbf{q}'}c_{\mathbf{k}'}|N_{\mathbf{k+q}}, N_{\mathbf{k}}; n_{\mathbf{q}}\rangle. \quad (2.51)$$

The summation gives a non-zero result when $\mathbf{q}' = -\mathbf{q}$ and $\mathbf{k}' = \mathbf{k}+\mathbf{q}$. By using this result in the Fermi golden rule, the transition rate takes the form

$$W(\mathbf{k}+\mathbf{q} \to \mathbf{k};\mathbf{q}) = \frac{4\pi}{\hbar}|M_{\mathbf{q}}|^2(1-N_{\mathbf{k}})N_{\mathbf{k+q}}(n_{\mathbf{q}}+1)\delta(E_{\mathbf{k}}+\hbar\omega_{\mathbf{q}}-E_{\mathbf{k+q}}), \quad (2.52)$$

where we used the standard relations: $a^{\dagger}_{\mathbf{q}}|n_{\mathbf{q}}\rangle = \sqrt{n_{\mathbf{q}}+1}|n_{\mathbf{q}}+1\rangle$, $\quad a_{\mathbf{q}}|n_{\mathbf{q}}\rangle = \sqrt{n_{\mathbf{q}}}|n_{\mathbf{q}}-1\rangle$ etc. [10], and extra factor of 2 comes from the spin summation. On exactly similar lines, the other transition rates can be computed. In the above equation the numbers $N_{\mathbf{k}}$ and $n_{\mathbf{q}}$ represent the number of electrons and the number of phonons in their respective states, in a given member of a statistical ensemble. When the statistical ensemble average is performed, these numbers change to mean occupation numbers. Let us define $\langle N_{\mathbf{k}}\rangle \equiv g_{\mathbf{k}}$ and $\langle n_{\mathbf{q}}\rangle \equiv n_{\mathbf{q}}$, where g_k and n_q are ensemble averaged distributions. On collecting all the above information, the collision term takes the form:

$$\begin{aligned}
\left(\frac{dg_{\mathbf{k}}}{dt}\right)_{coll} &= \frac{4\pi}{\hbar}\int \frac{d^3q}{(2\pi)^3}|M_{\mathbf{q}}|^2\{g_{\mathbf{k+q}}(1-g_{\mathbf{k}})[(n_{\mathbf{q}}+1)\delta(E_{\mathbf{k}}+\hbar\omega_q - E_{\mathbf{k+q}}) \\
&+ n_{-\mathbf{q}}\delta(E_{\mathbf{k}} - E_{\mathbf{k+q}} - \hbar\omega_{-\mathbf{q}})] - g_{\mathbf{k}}(1-g_{\mathbf{k+q}}) \\
&\times [(n_{-\mathbf{q}}+1)\delta(E_{\mathbf{k+q}} + \hbar\omega_{-q} - E_{\mathbf{k}}) + n_{\mathbf{q}}\delta(E_{\mathbf{k+q}} - E_{\mathbf{k}} - \hbar\omega_{\mathbf{q}})]\}
\end{aligned} \quad (2.53)$$

To save writing we will write k, q instead of $\mathbf{k}$, $\mathbf{q}$ (but keep in mind that k, q are vectors). For further simplifications, define $W_q^0 = \frac{4\pi}{\hbar}|M_{\mathbf{q}}|^2$, and define $U(k+q,k) = W_q^0[(n_q+1)\delta(E_{k+q} - E_k - \hbar\omega_q) + n_{-q}\delta(E_{k+q} - E_k + \hbar\omega_{-q})]$, and $U(k, k+q) = W_q^0[(n_{-q}+1)\delta(E_{k+q} - E_k + \hbar\omega_{-q}) + n_q\delta(E_{k+q} - E_k - \hbar\omega_q)]$. On implementing these changes, the collision integral can be written in symmetrical form

$$\left(\frac{dg_{\mathbf{k}}}{dt}\right)_{coll} = \int \frac{d^3q}{(2\pi)^3}\{U(k+q,k)g_{k+q}(1-g_k) - U(k,k+q)g_k(1-g_{k+q})\}. \tag{2.54}$$

To simplify further analysis, observe the following relationship between $U(k+q,k)$ and $U(k,k+q)$. Substitute the expression for the phonon distribution function (Equation (2.44)) in $U(k+q,k)$ and use the property $(f(x)\delta(x-a) = f(a)\delta(x-a))$ of the delta function in the first term in U. One can see that

$$U(k+q,k) = e^{\beta E_{k+q}}e^{-\beta E_k}U(k,k+q), \text{ or } e^{-\beta E_{k+q}}U(k+q,k) = e^{-\beta E_k}U(k,k+q). \tag{2.55}$$

We will use this relation in subsequent analysis. But before that, we linearize the Bloch-Boltzmann equation as we are discussing a near equilibrium situation which is induced by weak fields.[†††] Let us write

$$g_k = f_k + \delta g_k, \tag{2.56}$$

where δg_k is a tiny perturbation on the equilibrium distribution function f_k. On substituting the above expression into Equation (2.54), and retaining linear terms in δg_k:

$$\begin{aligned}\left(\frac{dg_{\mathbf{k}}}{dt}\right)_{coll} &= \int \frac{d^3q}{(2\pi)^3}\{\delta g_{k+q}[U(k+q,k)(1-f_k) + U(k,k+q)f_k] \\ &- \delta g_k[U(k+q,k)f_{k+q} + U(k,k+q)(1-f_{k+q})]\}.\end{aligned} \tag{2.57}$$

To obtain the above equation one must remember that in equilibrium the collision integral must be zero, so all the equilibrium terms will vanish. To further simplify it is convenient to write

$$\delta g_k = \beta f_k(1-f_k)\delta\Phi(k). \tag{2.58}$$

[†††]This is quite reasonable at ordinary field strengths where Ohm's law is valid.

And write $U(k+q,k) = U(k,k+q)e^{-\beta(E_k - E_{k+q})}$ as we have shown above. On incorporating the above substitutions into Equation (2.57) and simplifying further using the explicit expressions of f_k and f_{k+q}, one obtains

$$\left(\frac{dg_{\mathbf{k}}}{dt}\right)_{coll} = \beta \int \frac{d^3q}{(2\pi)^3}\{U(k,k+q)f_k(1-f_{k+q})(\delta\Phi(k+q)-\delta\Phi(k))\}. \quad (2.59)$$

Now, on substituting the expression of $U(k,k+q)$ (which is given in the paragraph above Equation (2.54)) and again using the property $f(x)\delta(x-a) = f(a)\delta(x-a)$ of the delta function, one can show, with straightforward algebra, that

$$\begin{aligned}\left(\frac{dg_{\mathbf{k}}}{dt}\right)_{coll} &= \beta \int \frac{d^3q}{(2\pi)^3} W_q^0 n_q \{f_{k+q}(1-f_k)\delta(E_{k+q} - E_k + \hbar\omega_{-q}) \\ &+ f_k(1-f_{k+q})\delta(E_{k+q} - E_k - \hbar\omega_q)\} \\ &\times [\delta\Phi(k+q) - \delta\Phi(k)]. \end{aligned} \quad (2.60)$$

With this, the Boltzmann equation (2.46) (now the Bloch-Boltzmann equation because of the electron-phonon collision term!) can be written as

$$\begin{aligned} &- \frac{e}{\hbar}\mathscr{E}_x v_x \frac{\partial f_k}{\partial E_k} = \left(\frac{dg_{\mathbf{k}}}{dt}\right)_{coll} = \beta \int \frac{d^3q}{(2\pi)^3} W_q^0 n_q \{f_{k+q}(1-f_k) \\ &\times \delta(E_{k+q} - E_k + \hbar\omega_{-q}) + f_k(1-f_{k+q})\delta(E_{k+q} - E_k - \hbar\omega_q)\} \\ &\times [\delta\Phi(k+q) - \delta\Phi(k)], \end{aligned} \quad (2.61)$$

where we consider that the electric field is applied in the x-direction (since we are dealing with an isotropic case, it is not an issue), and as we are dealing with the case where there is no temperature gradient, thus no space dependence of the distribution function is possible if we further ignore higher order thermoelectric effects. Since we are dealing with linearized Boltzmann equation and the nonequilibrium distribution function is of the first order in the applied electric field, we replace g_k by f_k in the drift term of the above equation. v_x in the above equation is the electron velocity in the $x-$direction. Equation (2.61) is the central result, and is called the Bloch-Boltzmann equation.

Our next aim is to explicitly evaluate the collision term in the above Bloch-Boltzmann equation. It turns out that full analytic solution at all temperatures is not possible. However, at very low and at very high temperatures the collision integral can be evaluated and its temperature dependence can be inferred. From the approximated collision integral the nonequilibrium distribution function will be evaluated and the conductivity will be evaluated from Ohm's law.

To evaluate the collision integral first we need to find the range of the phonon wave-vector q. We are going to assume the Debye model for phonons and impose an upper cut-off on q: $0 < q < q_D$ or $0 < \hbar\omega < \hbar\omega_D = k_B\Theta_D$, where Θ_D is the Debye temperature. Since the calculation is done for electron-phonon scattering in metals it is required to quantitatively compare the limits of q with the typical electron wave-vector k_F. Electrons that take part in conduction (in the diffusion zone of width $\sim k_BT$ around the Fermi surface, where $k_BT << \mu$) have wave-vectors $k \simeq k_F$ and $q << k_F$. But there could be another possibility that we have to rule out in the case of metals.

It turns out that the typical phonon energies are very small ($\sim meV$) as compared to the typical electronic energies ($\sim eV$). Due to this fact another condition is possible. By energy conservation:

$$\begin{aligned} E_{k\pm q} &= E_k + \hbar\omega_{\pm q}, \\ E_{k\pm q} &\simeq E_k, \\ (k\pm q)^2 &\simeq k^2, \\ \text{either,}\ q &<< k, \\ \text{or,}\quad \pm\mathbf{q}(2\mathbf{k}\pm\mathbf{q}) &\simeq 0\ , \\ \text{which implies}\ \mathbf{q} &\simeq \pm 2\mathbf{k}. \end{aligned} \tag{2.62}$$

We can rule out that the last condition $\mathbf{q} \simeq \pm 2\mathbf{k}$ is not relevant for metals (it can be relevant for semiconductors). To rule out, write in magnitudes

$$\begin{aligned} q &\simeq 2k, \\ \hbar s_l q &\simeq 2\hbar s_l k, \\ \hbar\omega_q &\simeq 2\left(\frac{k}{q_D}\right) k_B\Theta_D, \\ \hbar\omega_q &\simeq \sqrt{\frac{4\mu}{D}} k_B\Theta_D, \end{aligned} \tag{2.63}$$

where we defined $k \simeq \frac{\sqrt{2m\mu}}{\hbar}$ and $q_D = \frac{\sqrt{2mD}}{\hbar}$. The above condition is violated for metals as due to high carrier density: $\mu >> D$. But it can be true in the case of semiconductors in which due to low carrier density $\mu \simeq D$. Since we do not consider the case of semiconductors we will not use this condition, instead for metals we are going to use $0 < q < q_D$.

Let us return back to our main aim of the evaluation of the collision integral. We are going to simplify it using an iterative procedure similar to one used in the previous section. If we observe the nonequilibrium distribution function under RTA (refer to Equation (2.16))

$$\delta g_k = -\tau e v_x \mathscr{E}(-\frac{\partial f_k}{\partial E_k}), \tag{2.64}$$

then we notice that δg_k is proportional, along with other factors, to the relaxation time τ. Now, if the relaxation time is energy dependent, then δg_k will be proportional to the product of k_x (say the field is along x-direction) and some function of energy say $c(E_k)$. Thus,

$$\delta g_k = -k_x c(E_k)\left(-\frac{\partial f_k}{\partial E_k}\right). \tag{2.65}$$

In view of Equation (2.58), we notice that $\delta\Phi(k) = -k_x c(E_k)$, as we consider spherical Fermi surface $E_k = \hbar^2 k^2/(2m)$. On writing Φ in this manner in the Bloch-Boltzmann equation (2.61), using the property $f(x)\delta(x-a) = f(a)\delta(x-a)$ of the delta functions, and writing the expression for W_q^0, the

Bloch-Boltzmann equation takes the form

$$
\begin{aligned}
& \mathscr{E}_x k_x \frac{\partial f_k}{\partial E} = \\
& A\beta \int d^3q \frac{q^2}{\omega_q} n_q \{ f(E - \hbar\omega_q)(1 - f(E))\delta(E_{k+q} - E + \hbar\omega_q) \\
\times \quad & [(k_x + q_x)c(E - \hbar\omega) - k_x c(E)] + f(E)(1 - f(E + \hbar\omega_q)) \\
\times \quad & \delta(E_{k+q} - E - \hbar\omega_q)[(k_x + q_x)c(E + \hbar\omega_q) - k_x c(E)]\}, \qquad (2.66)
\end{aligned}
$$

where we lump all the constants into a constant A. As we are interested in the nature of temperature dependence only, we will keep adding all the irrelevant constants into it. Also we use E instead of E_k to save writing, and it is clear from the context. As ω_q is an even function of q, ω_{-q} is written as ω_q.

Our next step is to perform the integration over q:

$$
\int d^3q = \int_0^{q_D} dq q^2 \int_0^{\pi} d\vartheta \sin\vartheta \int_0^{2\pi} d\phi, \qquad (2.67)
$$

where the polar axis is along the direction of $\mathbf{k}$, ϑ is the angle between $\mathbf{q}$ and $\mathbf{k}$, and ϕ is the azimuth angle. Further assume that $\mathbf{k}$ and $\mathbf{q}$ subtends angles θ and θ_1 with $x - axis$, respectively. Refer to Figure 2.4. The component q_x can be

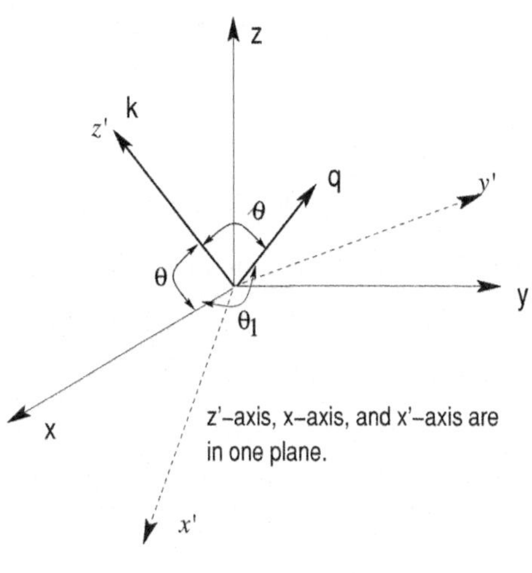

FIGURE 2.4: Diagram for the evaluation of q integral. Electric field is along the $x-$direction.

written in two ways: $q_x = q\cos\theta_1$, and also $q_x = q(\cos\vartheta\cos\theta + \sin\vartheta\cos\phi\sin\theta)$. In the second expression, $\mathbf{q}$ is resolved along $\mathbf{k}-$axis or $z'-$axis and $x'-$axis (refer to Figure 2.4). As $z'-$axis, $x-$axis, and $x'-$axis can be made to be in one plane (with angle $90 - \theta$ between $x'-$ and $x-$axis), q_x can be written as in the second expression, i.e., by further resolving along the $x-$axis (refer to Figure 2.4). Integration over the azimuth angle can be performed easily

$$\int_0^{2\pi} d\phi q_x = q\int_0^{2\pi} d\phi(\cos\vartheta\cos\theta + \sin\vartheta\cos\phi\sin\theta) = 2\pi q\cos\vartheta(k_x/k). \quad (2.68)$$

The value of $\cos\vartheta$ can be obtained from energy conservation constraint:

$$\begin{aligned} E_{k+q} - E_k \pm \hbar\omega_q &= 0 \\ q^2 + 2kq\cos\vartheta \pm \frac{2m}{\hbar}\omega_q &= 0 \\ \cos\vartheta &= -\left(\frac{q}{2k} \pm \frac{ms_l}{k}\right). \end{aligned} \quad (2.69)$$

Thus,

$$\int_0^{2\pi} d\phi q_x = -\pi k_x\left(\frac{q^2}{k^2} \pm \frac{\hbar\omega_q}{E}\right) \quad (2.70)$$

Next, the integration over the angle ϑ can be performed as follows:

$$\begin{aligned} I_{k,q} &\equiv \int_0^{\pi} d\vartheta\sin\vartheta\delta\left(\frac{\hbar^2}{2m}(\mathbf{k}+\mathbf{q})^2 - \frac{\hbar^2k^2}{2m} \pm \hbar\omega_q\right) \\ &= \frac{m}{\hbar^2kq}\int_0^{\pi} d\vartheta\sin\vartheta\delta(\cos\vartheta + \frac{q}{2k} \pm \frac{ms_l}{\hbar k}). \end{aligned} \quad (2.71)$$

With the substitution $x = \cos\vartheta$ the last integral can be easily evaluated with result $\frac{m}{\hbar^2kq}$, as phonon momentum, is much less than the electron momentum, i.e., $ms_l << \hbar k$ and also $q << k$ (good in metals). On collecting this information, and writing $\beta f(E)(1 - f(E)) = -\frac{\partial f}{\partial E}$, the collision integral of the Bloch-Boltzmann equation takes the form:

$$\begin{aligned} \left(\frac{dg_{\mathbf{k}}}{dt}\right)_{coll} &= -A\frac{k_x}{k}\frac{\partial f}{\partial E}\int_0^{q_D} dq\frac{q^3}{\omega_q}n_q\left(\frac{f(E-\hbar\omega_q)}{f(E)}\right. \\ \times &\quad [\{1 - q^2/(2k^2) - \hbar\omega_q/(2E)\}c(E-\hbar\omega_q) - c(E)] \\ + &\quad \left. e^{\beta\hbar\omega_q}\frac{f(E+\hbar\omega_q)}{f(E)}[\{1 - q^2/(2k^2) + \hbar\omega_q/(2E)\}c(E+\hbar\omega_q) - c(E)]\right). \end{aligned} \quad (2.72)$$

Introduce dimensionless variables $\eta = \beta(E - \mu)$ and $z = \beta\hbar\omega_q$, and introduce $D = \frac{\hbar^2 q_D^2}{2m}$. Both terms in the integrand are symmetric to $z \to -z$ if we replace $\frac{1}{e^z-1}$ by $\frac{1}{|e^z-1|}$. Thus, the integrand becomes an even function in z,‡‡‡ and the collision integral can be written in a compact form

$$\begin{aligned}\left(\frac{dg_{\mathbf{k}}}{dt}\right)_{coll} &= -AE^{-3/2}k_x\frac{\partial f}{\partial E}\left(\frac{T}{\Theta_D}\right)^3 \\ &\times \int_{-\frac{\Theta_D}{T}}^{+\frac{\Theta_D}{T}} dz\frac{z^2}{|1-e^{-z}|}\frac{e^\eta+1}{e^{\eta+z}+1} \\ &\times [(E + k_BTz/2 - D(T/\Theta_D)^2z^2/2)c(\eta+z) - Ec(\eta)].\end{aligned} \tag{2.73}$$

This leads to an integral equation in $c(z)$ whose general analytical solution is not possible. In the high temperature limit $T >> \Theta_D$, however, the above collision term reduces to the form of the relaxation time approximation.

Let us perform the high temperature expansion of $c(z)$. In the high temperature limit, as $\frac{\Theta_D}{T} << 1$, the limits of the integration are very small as compared to one, and thus $z << 1$. Expand $c(\eta+z) \simeq c(\eta) + z\frac{\partial c(\eta)}{\partial \eta} + ...$ The leading order term in the parenthesis is the one with D, this leads to

$$\left(\frac{dg_{\mathbf{k}}}{dt}\right)_{coll} = ADc(\eta)E^{-3/2}k_x\frac{\partial f}{\partial E}\left(\frac{T}{\Theta_D}\right)^5\int_{-\frac{\Theta_D}{T}}^{+\frac{\Theta_D}{T}} dz\frac{z^4}{|1-e^{-z}|}\frac{e^\eta+1}{e^{\eta+z}+1}. \tag{2.74}$$

Although $T >> \Theta_D$, but $k_BT << \mu$. Thus $-\frac{\partial f}{\partial E}$ is still like a delta function around the Fermi energy, thus $\eta \simeq 0$ (use the property of the delta function). Now $g_k = f_k + \beta f_k(1 - f_k)k_xc(\eta) = f_k - \frac{\partial f_k}{\partial E}k_xc(\eta)$, we obtain:

$$\left(\frac{dg_{\mathbf{k}}}{dt}\right)_{coll} = -\frac{g_k - f_k}{\tau(T)}, \tag{2.75}$$

which is nothing but the collision term in the relaxation time approximation with

$$\frac{1}{\tau(T)} = A\mu^{-3/2}\left(\frac{T}{\Theta_D}\right)^5\int_{-\frac{\Theta_D}{T}}^{+\frac{\Theta_D}{T}} dz\frac{z^4}{|1-e^{-z}|(e^z+1)}. \tag{2.76}$$

‡‡‡As z runs over positive values in the original integrand, this change is permissible.

In the limit under consideration the integral behaves as $\sim \left(\frac{\Theta_D}{T}\right)^4$, thus

$$\frac{1}{\tau(T)} \sim T. \tag{2.77}$$

We had already solved the Boltzmann equation in RTA and found $\rho = \frac{m}{ne^2}\frac{1}{\tau(T)}$. Thus we obtain our final result: $\rho \propto T$ when $T >> \Theta_D$, which is well reproduced by experiments! It is quite surprising that one could obtain a reasonable result from the Bloch-Boltzmann equation which is solved under a large set of assumptions. This only shows that the assumptions made are reasonable, at least for the case of simple metals, and the agreement between the theory and experiment is good in the restricted temperature range considered above.

2.4.1 The Grüneisen formula

Our next aim is to obtain an approximated solution of the Bloch-Boltzmann equation in the low temperature limit $T << \theta_D$ or more appropriately in somewhat general form. It is possible to proceed using an iterative procedure that works well when the Fermi energy is the highest energy scale in the problem. We have obtained:

$$\begin{aligned} \mathscr{E}_x \frac{\partial f}{\partial E} &= -AE^{-3/2} k_x \frac{\partial f}{\partial E}\left(\frac{T}{\Theta_D}\right)^3 \\ &\times \int_{-\frac{\Theta_D}{T}}^{+\frac{\Theta_D}{T}} dz \frac{z^2}{|1-e^{-z}|} \frac{e^\eta + 1}{e^{\eta+z}+1} \\ &\times [(E + k_B T z/2 - D(T/\Theta_D)^2 z^2/2)c(\eta + z) - Ec(\eta)]. \end{aligned} \tag{2.78}$$

Now write the above equation in the following way:

$$\begin{aligned} A^{-1} & \mathscr{E}_x E^{3/2} \left(\frac{\Theta_D}{T}\right)^3 \frac{\partial f}{\partial E} \\ + & \frac{\partial f}{\partial E} \int_{-\frac{\Theta_D}{T}}^{+\frac{\Theta_D}{T}} dz z^2 (k_B T z/2 - D(T/\Theta_D)^2 z^2/2) c(\eta + z) \frac{e^\eta + 1}{(e^{\eta+z})|1-e^{-z}|} \\ = & E\frac{\partial f}{\partial E} \int_{-\frac{\Theta_D}{T}}^{+\frac{\Theta_D}{T}} dz z^2 (c(\eta) - c(\eta+z)) \frac{e^\eta + 1}{(e^{\eta+z})|1-e^{-z}|}. \end{aligned} \tag{2.79}$$

We will attempt its solution via an iterative procedure (refer for details [11]). A homogeneous equation is obtained by neglecting the LHS which will have a solution of the form $c(\eta) = constant = \alpha$. Now the non-homogeneous equation (the full equation) can only be solved if the LHS is orthogonal to the solution of the homogeneous equation (the basic requirement for any legal solution of the full equation). Thus,

$$\begin{aligned} & A^{-1}\mathscr{E}_x \int_{-\infty}^{+\infty} d\eta E^{3/2}\left(\frac{\Theta_D}{T}\right)^3 \frac{\partial f}{\partial E} \\ = & -\int_{-\infty}^{+\infty} d\eta \frac{\partial f}{\partial E} \int_{-\frac{\Theta_D}{T}}^{+\frac{\Theta_D}{T}} dz z^2 \left(\frac{k_B T z}{2} - D(T/\Theta_D)^2 \frac{z^2}{2}\right)[\alpha] \frac{e^\eta + 1}{(e^{\eta+z})|1-e^{-z}|} \end{aligned} \tag{2.80}$$

from which α can be calculated. Our next step is to investigate for a more accurate solution. Now, if the operator on the LHS of Equation (2.79) is small, then a further improvement on the constant solution can be obtained by setting $c(\eta) = \alpha + \beta(\eta)$, and $\beta(\eta)$ can be obtained by solving the original Equation (2.79). It turns out that the operator on the left of Equation (2.79) is small, it yields small $\beta(\eta) << \alpha$[11]. Restricting to the leading order yields

$$\alpha = \frac{A^{-1}\mathscr{E}_x \mu^{-3/2}(\Theta_D/T)^5}{D\int_0^{\Theta_D/T} dz \frac{z^5}{(e^z-1)(1-e^{-z})}} = A^{-1}\mathscr{E}_x \mu^{3/2}(\Theta_D/T)^5 J_5^{-1}(\Theta_D/T), \tag{2.81}$$

where we have used the fact that $D >> k_B\Theta_D$ and the term with D is the dominating term. The integration over η can be easily performed (refer to problem 6 in the last chapter). Thus, from Equation (2.65)

$$\delta g_k = -k_x \alpha(-\partial f/\partial E), \tag{2.82}$$

and the current density is given by the standard expression: $J_x = -2e\int \frac{d^3k}{(2\pi)^3} v_k^x \delta g_k$ from which the conductivity, using $J_x = \sigma \mathscr{E}_x$, can be computed. The temperature-dependence of resistivity (1/conductivity) can be inferred (keeping only temperature-dependent factors):

$$\rho \propto \left(\frac{T}{\Theta_D}\right)^5 J_5 \left(\frac{\Theta_D}{T}\right). \tag{2.83}$$

This is the famous Grüneisen formula. The Debye function $J_5(x)$ has the following asymptotic behavior $J_5(x) \sim x^4$ when $x << 1$, and $J_5(x) \sim constant$ when $x >> 1$. Using this information we obtain what we set out to obtain:

$$\rho(T) \sim \begin{cases} T & : T >> \Theta_D \\ T^5 & : T << \Theta_D. \end{cases} \tag{2.84}$$

2.5 Summary points

1. We started this chapter by considering the premises on which the Boltzmann equation is justified when applied to electrons in metals. There is no serious difficulty when the Boltzmann equation is applied to molecules in a rarefied gas, as the system can be treated within classical mechanics. The quantum character of the problem of the dynamics of electrons in metals does not warrant a direct application. We need to justify the semiclassical model on which the application of the Boltzmann equation is based. Although the justification is "a-posteriori", some rough arguments in favor of the semiclassical model are advanced (Section 2.1).

2. Then we introduced the Chamber's method of the Boltzmann kinetic equation (the method of "tube integrals") and we showed that it corresponds to the relaxation time approximation in the full Boltzmann equation. DC resistivity is computed within spherical Fermi surface model and it agrees with the Drude formula.

3. Next we introduced the full Boltzmann equation (i.e., beyond relaxation time approximation). We notice that it respects the Pauli exclusion principle and it is not restricted to the dilute limit where mean occupation of a state is much less than one. From here, we reanalyzed the assumptions made in the RTA Boltzmann equation and observe that RTA form can be obtained under a set of assumptions which are collected in Section 2.3.1, namely, existence of a universal scattering rate; weak deviations from equilibrium; dilute limit; and elastic scattering.

4. Finally, we introduced the famous Bloch-Boltzmann theory. This theory resolves the problems of the Drude-Lorentz theory in addressing the temperature dependence of thermal and electrical conductivities. We considered the case of electrical resistivity in detail.

Bibliography

[1] N. Ashcroft and N. Mermin, *Solid State Physics*, Brooks-Cole (1976).

[2] Navinder Singh, arXive: 1103.4003 (2011).

[3] R. G. Chambers, *Proc. Phys. Soc.* **A65**, 458 (1952).

[4] C. MacCallum, *Phys. Rev.* **132**, 930 (1963).

[5] A. Lodder, *Physica*, **47**, 137 (1970).

[6] V. Heine, *Phys. Rev.* **107**, 431 (1957).

[7] H. Budd, *Phys. Rev.* **127**, 4 (1962).

[8] Navinder Singh, *Int. J. Mod. Phys.* B. **24**, 1141 (2010).

[9] G. D. Mahan, *Many-Particle Physics*, 3rd edition, Springer (2000).

[10] H. Bruus and K. Flensberg, *Many-Body Quantum Theory in Condensed Matter Physics: An Introduction*, Oxford Graduate Texts (2004).

[11] A. H. Wilson, *The Theory of Metals*, CUP (1953).

Chapter 3

Some techniques from nonequilibrium statistical mechanics

Brownian motion is not merely random motion of a very fine particle; in general it is random motion of a physical quantity to be observed in a macrosystem

M. Toda, R. Kubo, and N. Saito *.

In the previous chapter the traditional method of the Boltzmann equation was introduced and it was shown that the Boltzmann equation leads to the Drude formula under RTA. The full Boltzmann equation or the "beyond RTA" Boltzmann equation was also introduced. The Bloch-Boltzmann theory and the case of electron-phonon scattering in DC resistivity was explicitly considered. One has to carefully analyze the premises on which the Boltzmann equation, as applied to the electrons in metals, is based. We heuristically justified that using the uncertainty principle and recognizing that electrons can be treated as wave packets with size much greater than a lattice constant† and thus a semiclassical distribution function $g(\mathbf{r}, \mathbf{k}, t)$ was written. The wave vector $\mathbf{k}$ remains a good quantum number as an uncertainty in it (arising due to finite mean free path that is further due to scattering) remains small as compared to its magnitude ($\Delta k << |\mathbf{k}| \sim k_F$).

* *Statistical Physics II: Nonequilibrium Statistical Mechanics*, Springer (2000).

†In a perfect lattice, electron wave function is a Bloch wave (a modulated plane wave) of macroscopic extent (of a size of the sample).

Now compare and contrast the above situation with a situation in which strong electron-electron interaction or any other strong interaction is present such that momentum is no longer a good quantum number. Consider a situation in which the mean free path is of the order of lattice constant (a). Then the uncertainty in momentum will be of the order of the reciprocal wave vector ($\Delta k \sim \frac{1}{a}$) which is of the order of $\mathbf{k}$ itself. In such a situation $g(\mathbf{r}, \mathbf{k}, t)$ itself cannot be defined and the Boltzmann equation has no justification for such a case.

The purpose of the present chapter is to introduce a general method that is "fully quantum" in nature. Thus, it goes beyond the validity of the Boltzmann equation and in appropriate limits it leads to the results that can be derived using the Boltzmann equation. The method is much more general in its physical content and offers a more general mathematical apparatus to deal with transport problems (i.e., the methods of quantum field theory in condensed matter physics).

However, this method has its limitations. It cannot be applied to situations which are arbitrarily far from equilibrium. It works in the linear regime. The method is called the linear response theory, and Kubo's formula relating transport coefficients with time correlation functions is a central piece in the whole formalism. The other important topics (in fact, alternative formulations) are (1) Zwanzig's projection operator techniques and (2) Mori's formulation of Kubo's theory using these projection operators which are discussed in the next chapter.

In the present chapter we discuss the Kubo formalism in an elementary way (almost all the mathematical steps are given). Our focus will remain on the problem of electrical conduction.

To provide a coherent picture of this beautiful field, we start in the next section with Nyquist's thermodynamical arguments pertaining to the Fluctuation-Dissipation Theorem (FDT). FDT is the queen of nonequilibrium statistical mechanics. Its status is like that of the partition function of equilibrium statistical mechanics. Once the partition function is known, all the required equilibrium observable can be computed. Similarly, by exploiting the FDT, transport coefficients can be calculated. We will see how Kubo's formulation provides the "complete picture."

3.1 Nyquist's thermodynamical arguments and Callen-Welton's fluctuation-dissipation theorem

If one connects a resistor with a sensitive voltmeter, one observes an AC (noise) voltage of zero mean but non-zero Root Mean Square (RMS) value of very small magnitude (of sub-micro volts in ordinary conductors). This thermally induced noise voltage (called the Johnson-Nyquist(JN) noise) was first observed by John B. Johnson of Bell Labs in 1926 and its theoretical explanation was given by his colleague Harry Nyquist in 1927. Nyquist proved that the frequency integrated variance of voltage is proportional to the frequency integrated resistance,

$$\langle V^2 \rangle = 4k_B T \int_0^\infty R(\omega) d\omega. \tag{3.1}$$

Below we collect Nyquist's thermodynamical arguments which lead to the above equation. His arguments are simple but ingenious [1]. He first shows that if one connects two *equal* resistors in parallel with each other (Figure 3.1(a)), then the power dissipated in the first resistor due to JN noise voltage produced by the second is equal to the power dissipated in the second due to JN noise produced by the first, provided both resistors are kept at the same temperature. Thus, there is a balance of power flow. This is in accordance with the second law of thermodynamics (no spontaneous heating of the either resistor).

He further refines his argument by showing that the above inference is true

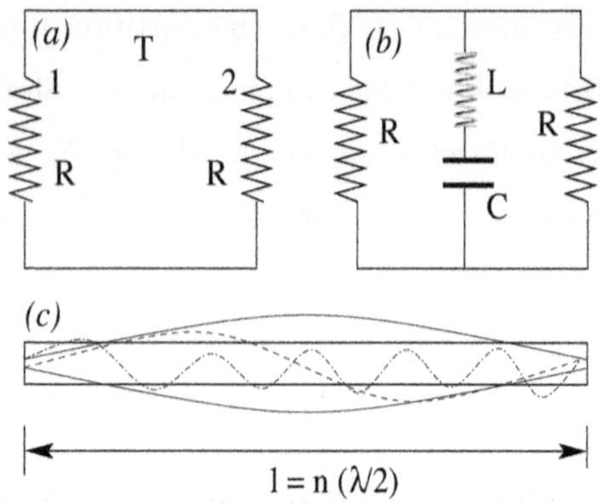

FIGURE 3.1: (a) Two resistors, both at the same temperature, are connected with each other. (b) An arrangement to show that in any frequency interval equal power flows from both sides (power flow balance). (c) Standing modes in a shunted transmission line.

in all frequency intervals. It is proved by using "proof by contradiction." LC circuit is connected in between the resistors (Figure 3.1(b)) which acts like a shunt at a specific resonance frequency $\omega = \frac{1}{\sqrt{LC}}$. Let us assume that the power transfer is nonuniform in frequency, say, more power flows from resistor 1 to resistor 2 in the frequency interval ω to $\omega + d\omega$ and to compensate this equivalent power flows from resistor 2 to resistor 1 in the interval ω' to $\omega' + d\omega'$. Now due to LC resonant circuit more power that flows from 1 to 2 will be shunted (as compared to that from 2 to 1). But the power flow from 2 to 1 will not be shunted (as LC is not resonant at ω'). Thus, the arrangement in Figure 3.1(b) will spontaneously lead to the heating of resistor number 1 as compared to the other, again violating the second law of thermodynamics.

Thus he concludes that there is a balance of power flow in each frequency interval and power transferred must be a universal function of frequency. Further, it was known from the experiments of Johnson that the JN noise is a universal function of resistance and temperature (it does not depend on the material of the resistor). Thus $V_{rms} = f(R, T, \omega)$, where f is some universal function. Nyquist explicitly derives this universal function using the law of equipartition of energy as given next.

In his thought experiment he first disconnects the two resistors and then connects them at the ends of a long lossless transmission line. This connection lasts for a time interval l/v, where l is the length of the transmission line, and v is the speed of electromagnetic waves in the line. After this, he again disconnects the resistors from the line, and short circuits the ends of the line. The transmission line acts like a cavity and standing modes step up (Figure 3.1(c)).

To compute the power transmitted from the resistors to the transmission line he calculates the mode density as follows. The number of standing modes in the frequency interval f to $f+df$ will be $(2l/v)df$ (ends must be nodes). By equipartition of energy each mode has k_BT of energy (in the classical regime $hf << k_BT$). Thus, the energy in the frequency interval f to $f+df$ will be $k_BT(2l/v)df$. The average power transferred in time interval l/v is $2k_bTdf$ (energy/time). This can be connected with voltage. In a linear system (obeying Ohm's law) the average power is $\frac{\langle V^2\rangle_f}{2R}df$ (where $\langle V^2\rangle_f$ is the variance of voltage per unit frequency). Therefore, $\langle V^2\rangle_f df = 4k_BTRdf$. If the resistor (the real part of impedance) is frequency dependent, then the frequency integrated variance bears the relation

$$\langle V^2\rangle = 4k_BT\int_0^\infty R(\omega)d\omega, \tag{3.2}$$

which is known as the Nyquist theorem [1]. *The fundamental importance of this relation and other general linear fluctuation-dissipation theorems (see below) is that the equilibrium fluctuations (e.g., JN noise) have "hidden" information regarding transport coefficients (here the resistance $Re[Z(\omega)]$).* Thus, from the nature of equilibrium fluctuations the nature of transport coefficients can be inferred and vice versa. This fact is the cornerstone of linear nonequilibrium statistical mechanics [2]. Kubo exploited this connection and showed that, in a very general way, transport coefficients can be expressed as integrals of equilibrium correlation functions (see next). From this ensues an alternative approach to the Bloch-Boltzmann theory.

Before we present Kubo's landmark work, another important work by Callen and Welton (CW) is worth considering [3]. CW's work gave a rigorous quantum mechanical foundation to Nyquist's qualitative result and it is known as the Fluctuation-Dissipation Theorem (FDT).

To prove Equation (3.2) from more fundamental principles, CW considers a conductor of length L and of impedance $Z(\omega)$ biased with an AC voltage $V(t) = V_0 \sin(\omega t)$.[‡] The total Hamiltonian H of the system comprises two parts: the unperturbed part H_0, and the perturbation $V(t)\hat{Q}$. That is,

$$H = H_0 + V(t)\hat{Q}, \quad \text{where } \hat{Q} = \sum_i e\frac{\hat{x}_i}{L}, \tag{3.3}$$

where x_i is the position of the ith particle from one end of the conductor and $\hat{x}_i$ is the corresponding operator. They assume that eigenenergies E_n and eigenfunctions ϕ_n of H_0 are known.

Next, they compute average power absorbed by the conductor from the battery. This is done with perturbation theory. Let the wavefunction of the perturbed system be $\psi(t) = \sum_n a_n(t)\phi_n$ which obeys the Schroedinger equation, $i\hbar\frac{\partial\psi(t)}{\partial t} = (H_0 + V(t)\hat{Q})\psi(t)$. From the Schroedinger equation the expansion coefficients $a_n(t)$ can expressed as

$$\dot{a}_m(t) = -\frac{i}{\hbar}E_m a_m(t) - \frac{i}{\hbar}V_0 \sin(\omega t)\sum_n Q_{mn}a_n(t). \tag{3.4}$$

Here matrix element Q_{mn} stands for $\int dx\phi_m^*(x)\hat{Q}(x)\phi_n(x)$ as usual. The transformation $a_n(t) = b_n(t)e^{-\frac{i}{\hbar}E_n t}$ removes the first term on the right-hand side of the above equation. The transition probability $\mathscr{P}_{if}(t)$ to find the system in some final state $|f\rangle$ of energy E_f at time t when the system was prepared in an initial state $|i\rangle$ of energy E_i at an initial time $t = 0$ is given by $\mathscr{P}_{if}(t) = |\langle f|\psi(t)\rangle|^2 = |a_f(t)|^2 = |b_f(t)|^2$. Then by standard first order per-

‡Here, the electrical circuit case is an illustrative example. Applicability of the results is much more general [3].

turbation theory and by noticing that $b_m^{(0)}(t) = \delta_{mi}$ the transition probability can be written as

$$\begin{aligned} \mathscr{P}_{if}(t) &= \frac{V_0^2}{\hbar^2}|Q_{fi}|^2 \left| \int_0^t dt' e^{-i\omega_{if}t'} \sin(\omega t') \right|^2 \\ &= \frac{1}{2}\pi t \frac{V_0^2}{\hbar}|Q_{fi}|^2(\delta(\omega+\omega_{if}) + \delta(\omega - \omega_{if})). \end{aligned} \tag{3.5}$$

To prove the last equality in the above equation write $\sin(x) = \frac{e^{ix}-e^{-ix}}{2i}$ and then in the long time limit ($t >> \frac{1}{\omega_{if}}$) use the identity $lim_{t\to\infty}\left|\frac{e^{ixt}-1}{x}\right|^2 = 2\pi t\delta(x)$. The meaning of the long time limit here is that the internal time scales $\frac{1}{\omega_{if}}$ of the system are much smaller than the external time scale $t \sim \frac{1}{\omega}$ associated with the driving field $V_0 \sin(\omega t)$. The reason why we want the long time limit is that we are interested in the average power absorbed in the macroscopic time scale $t \sim \frac{1}{\omega}$ of the applied field.

If the final states form a continuum (generally true for a system in the thermodynamic limit, as is the case here), then the total transition probability per unit time from the initial state $|i\rangle$ to *any final state* is given by

$$\begin{aligned} \mathscr{P}_i &= \frac{1}{t}\sum_f \mathscr{P}_{if}(t) = \frac{1}{t}\int dE_f \rho(E_f)\mathscr{P}_{if} \\ &= \frac{\pi}{2}\frac{V_0^2}{\hbar}(|\langle E_i + \hbar\omega|\hat{Q}|E_i\rangle|^2\rho(E_i+\hbar\omega) + |\langle E_i - \hbar\omega|\hat{Q}|E_i\rangle|^2\rho(E_i-\hbar\omega)). \end{aligned} \tag{3.6}$$

While replacing sum by integral we introduce the density of states $\rho(E)$. The two terms in the equality in the above equation (terms in the parenthesis) have the following physical meaning. The first term (i.e., $\langle E_i + \hbar\omega|\hat{Q}|E_i\rangle$) represents a physical process in which a photon is absorbed by the system (the conductor) from the battery and the initial state with energy E_i changes to a state with energy $E_i + \hbar\omega$ (photon absorption). The second term represents a loss of a single photon by the system to the environment (i.e., $\langle E_i - \hbar\omega|\hat{Q}|E_i\rangle$). Thus, the power absorbed by the conductor is

$$Power(E_i) = \frac{\pi\hbar\omega}{2}\frac{V_0^2}{\hbar}(|\langle E_i+\hbar\omega|\hat{Q}|E_i\rangle|^2\rho(E_i+\hbar\omega) - |\langle E_i-\hbar\omega|\hat{Q}|E_i\rangle|^2\rho(E_i-\hbar\omega)), \tag{3.7}$$

where $\hbar\omega$ is the energy of a single photon. Notice the minus sign between the two terms in the parenthesis. It represents gain minus loss. The *average* power absorbed at the deriving frequency ω is given by

$$\langle Power\rangle_\omega = \int_0^\infty dE_i \rho(E_i) f(E_i) Power(E_i). \tag{3.8}$$

It is obtained by taking the weighted summation over all possible initial states with the thermodynamic weighting factor $f(E_i) = e^{-E_i/k_B T}$.

For a system obeying Ohm's law (a linear system) the power is given as $\langle Power\rangle_\omega = \frac{1}{2}V_0^2 \frac{R(\omega)}{|Z(\omega)|^2} = \frac{1}{2}V_0^2 Y(\omega)$, where $Y(\omega)$ is the real part of the linear admittance $Y(\omega) = \frac{R(\omega)}{|Z(\omega)|^2}$. From the expression of the averaged power (Equation (3.8)) an expression for $Y(\omega)$ can be obtained

$$\begin{aligned} Y(\omega) &= \pi\omega \int_0^\infty dE \rho(E) f(E) \\ &\times (|\langle E+\hbar\omega|\hat{Q}|E\rangle|^2 \rho(E+\hbar\omega) - |\langle E-\hbar\omega|\hat{Q}|E\rangle|^2 \rho(E-\hbar\omega)). \end{aligned} \tag{3.9}$$

Next, Callen and Welton analyze the nature of equilibrium voltage fluctuations in the conductor (when the battery has been disconnected). The average current across the conductor $\langle\dot{\hat{Q}}\rangle = \frac{i}{\hbar}\langle n|[H_o, \hat{Q}]|n\rangle$ vanishes in equilibrium (as expected),[§] but its square does not. With a simple calculation they show that

$$\begin{aligned} \langle\dot{\hat{Q}}^2\rangle &= \int_0^\infty d\omega \hbar\omega^2 \int_0^\infty dE \rho(E) f(E) (|\langle E_i+\hbar\omega|\hat{Q}|E_i\rangle|^2 \rho(E_i+\hbar\omega) \\ &+ |\langle E_i-\hbar\omega|\hat{Q}|E_i\rangle|^2 \rho(E_i-\hbar\omega)). \end{aligned} \tag{3.10}$$

To prove the above equality proceed as follows. Insert a complete set of eigenstates of the unperturbed Hamiltonian in the expectation value of $\dot{\hat{Q}}^2$

$$\langle n|\dot{\hat{Q}}^2|n\rangle = \sum_m \langle n|\dot{\hat{Q}}|m\rangle\langle m|\dot{\hat{Q}}|n\rangle = -\frac{1}{\hbar^2}\sum_m \langle n|[H_0,\hat{Q}]|m\rangle\langle m|[H_0,\hat{Q}|n\rangle. \tag{3.11}$$

Using our notation for the matrix element the above expression can be simplified as

$$\langle n|\dot{\hat{Q}}^2|n\rangle = \frac{1}{\hbar^2}\sum_m (\hbar\omega_{nm})^2 |Q_{nm}|^2, \tag{3.12}$$

[§]Here, $|n\rangle$ is the eigenstate of the unperturbed Hamiltonian with eigen energy E_n.

where we have defined $\hbar\omega_{nm} = E_n - E_m$. To transform the above expression in a form that resembles the expression for admittance (Equation 3.9) divide the summation over m in two parts as $\sum_m = \sum_{m<n} + \sum_{m\geq n}$. Thus

$$\langle n|\dot{\hat{Q}}^2|n\rangle = \frac{1}{\hbar^2}\sum_{m<n}(\hbar\omega_{nm})^2|Q_{nm}|^2 + \frac{1}{\hbar^2}\sum_{m\geq n}(\hbar\omega_{nm})^2|Q_{nm}|^2. \tag{3.13}$$

Now, by converting the summations into integrals using Density of State (DOS) the above equation can be written as

$$\begin{aligned}\langle n|\dot{\hat{Q}}^2|n\rangle &= \frac{1}{\hbar^2}\int_0^{\infty}\hbar d\omega(\hbar\omega)^2|\langle E_n - \hbar\omega|\hat{Q}|E_n\rangle|^2\rho(E_n - \hbar\omega) \\ &+ \frac{1}{\hbar^2}\int_{-\infty}^{0}\hbar d\omega(\hbar\omega)^2|\langle E_n - \hbar\omega|\hat{Q}|E_n\rangle|^2\rho(E_n - \hbar\omega). \end{aligned} \tag{3.14}$$

Here the integration variables are changed from m to ω using $E_m = E_n - \hbar\omega$ and ω_{mn} is written as ω but positive and negative values are separated in the above equation. Next, by changing variables once more in the second integral from ω to $-\omega$ the above equation can be rewritten as

$$\begin{aligned}\langle n|\dot{\hat{Q}}^2|n\rangle &= \int_0^{\infty} d\omega\hbar\omega^2(|\langle E_n + \hbar\omega|\hat{Q}|E_n\rangle|^2\rho(E_n + \hbar\omega) \\ &+ |\langle E_n - \hbar\omega|\hat{Q}|E_n\rangle|^2\rho(E_n - \hbar\omega)). \end{aligned} \tag{3.15}$$

Above is the expectation value of the square of the current operator in the nth eigenstate. To compute the thermodynamical average one has to sum over all possible nth states (with the thermodynamical weighting factor). Thus the thermal expectation value of the square of the current operator is given by

$$\begin{aligned}&\langle\dot{\hat{Q}}^2\rangle = \sum_n f(E_n)\langle n|\dot{\hat{Q}}^2|n\rangle = \int_0^{\infty} dEf(E)\rho(E) \\ &\times\int_0^{\infty} d\omega\hbar\omega^2(|\langle E + \hbar\omega|\hat{Q}|E\rangle|^2\rho(E + \hbar\omega) + |\langle E - \hbar\omega|\hat{Q}|E\rangle|^2\rho(E - \hbar\omega)). \end{aligned} \tag{3.16}$$

Q.E.D.

This equation resembles that of the admittance (Equation (3.9)). This is the equal time correlation function for current. For the subsequent calculation we will also be needing the correlation function at different times, i.e.,

$\langle \dot{\hat{Q}}(t')\dot{\hat{Q}}(t'')\rangle$. The computation of it is similar to the computation of the equal time correlation function as given above. First, let us collect the time evolution in the last Heisenberg operator

$$\langle \dot{\hat{Q}}(t')\dot{\hat{Q}}(t'')\rangle = tr(e^{-\beta H_0} e^{iH_0 t'/\hbar}\dot{\hat{Q}}(0)e^{-iH_0 t'/\hbar}\dot{\hat{Q}}(t''))$$
$$= tr(e^{-\beta H_0}\dot{\hat{Q}}(0)e^{-iH_0 t'/\hbar}\dot{\hat{Q}}(t'')e^{+iH_0 t'/\hbar})$$
$$= tr(e^{-\beta H_0}\dot{\hat{Q}}(0)\dot{\hat{Q}}(t''-t')) = \sum_{m,n} e^{-\beta E_n}\langle n|\dot{\hat{Q}}(0)|m\rangle\langle m|\dot{\hat{Q}}(t''-t')|n\rangle. \quad (3.17)$$

Here the cyclic property of trace is used to shift the time evolution in the last operator. The rest of the calculation is the same as before. The summation over m is divided into two parts as done in Equation (3.13). Then by changing the summation into integration using DOS leads to

$$\begin{aligned}\langle \dot{\hat{Q}}(t')\dot{\hat{Q}}(t'')\rangle &= \int_0^\infty dE f(E)\rho(E)\int_0^\infty d\omega\hbar\omega^2(|\langle E+\hbar\omega|\hat{Q}|E\rangle|^2\rho(E+\hbar\omega)\\ &\times\; e^{i\omega(t''-t')} + |\langle E-\hbar\omega|\hat{Q}|E\rangle|^2\rho(E-\hbar\omega)e^{-i\omega(t''-t')}). \quad (3.18)\end{aligned}$$

Our aim is to establish the connection between the voltage fluctuations and the admittance. For this, first of all, we have to find a relation between current fluctuations and voltage fluctuations (i.e., between $\langle\dot{\hat{Q}}^2\rangle$ and $\langle V^2\rangle$¶). This connection is established using Ohm's law. But here there is a slight complication. The equilibrium current fluctuations are composed of all the harmonic components and we have to use Ohm's law in the frequency domain

$$v(\omega) = Z(\omega)\dot{q}(\omega). \quad (3.19)$$

$v(\omega)$ and $\dot{q}(\omega)$ are the Fourier transforms of $V(t)$ and $\dot{\hat{Q}}(t)$: $v(\omega) = \int_{-\infty}^{+\infty} dt V(t)e^{i\omega t}$, $V(t) = \int_{-\infty}^{+\infty}\frac{d\omega}{2\pi}v(\omega)e^{-i\omega t}$ and delta function is defined in the usual way $\delta(\omega) = \frac{1}{2\pi}\int_{-\infty}^{+\infty} dt e^{it\omega}$. Take the inverse Fourier transform of

¶The time index in $\langle\dot{\hat{Q}}^2(t)\rangle$ and in $\langle V^2(t)\rangle$ is not shown explicitly as these are equilibrium fluctuations and are independent of time.

Equation (3.19)

$$V(t) = \int_{-\infty}^{+\infty} \frac{d\omega}{2\pi} Z(\omega) e^{-i\omega t} \dot{q}(\omega) = \int_{-\infty}^{+\infty} \frac{d\omega}{2\pi} Z(\omega) \int_{-\infty}^{+\infty} dt' \dot{Q}(t') e^{i\omega t' - i\omega t}. \tag{3.20}$$

Square the above equation and take ensemble average (notice the dummy indices)

$$\begin{aligned} \langle V(t)^2 \rangle &= \int_{-\infty}^{+\infty} \frac{d\omega'}{2\pi} \int_{-\infty}^{+\infty} \frac{d\omega''}{2\pi} \int_{-\infty}^{+\infty} dt' \int_{-\infty}^{+\infty} dt'' Z(\omega') Z(\omega'') \\ &\times \langle \dot{Q}(t') \dot{Q}(t'') \rangle e^{i\omega' t' - i\omega' t} e^{i\omega'' t'' - i\omega'' t}. \end{aligned} \tag{3.21}$$

Plug the expression for the current-current correlation function (Equation (3.18)) into the above equation. The integrals over dt' and dt'' give delta functions $\delta(\omega + \omega'') = \frac{1}{2\pi} \int_{-\infty}^{+\infty} dt'' e^{it''(\omega+\omega'')}$ and $\delta(\omega + \omega') = \frac{1}{2\pi} \int_{-\infty}^{+\infty} dt' e^{it'(\omega'-\omega)}$ and the integrals over ω' and ω'' can be performed using the properties of the delta functions. This leads to

$$\begin{aligned} \langle V(t)^2 \rangle &= \int_0^{\infty} dE f(E) \rho(E) \int_0^{\infty} d\omega \hbar \omega^2 |Z(\omega)|^2 (|\langle E + \hbar\omega | \hat{Q} | E \rangle|^2 \rho(E + \hbar\omega) \\ &+ |\langle E - \hbar\omega | \hat{Q} | E \rangle|^2 \rho(E - \hbar\omega)). \end{aligned} \tag{3.22}$$

To establish the connection between Equation (3.9) and (3.22) consider the following integral which is common in both

$$I_{com} = \int_0^{\infty} dE f(E) \rho(E) |\langle E - \hbar\omega | \hat{Q} | E \rangle|^2 \rho(E - \hbar\omega). \tag{3.23}$$

Change variables, define $E' = E - \hbar\omega$. The above integral becomes

$$I_{com} = \int_0^{\infty} dE f(E + \hbar\omega) \rho(E + \hbar\omega) |\langle E | \hat{Q} | E + \hbar\omega \rangle|^2 \rho(E), \tag{3.24}$$

where the lower limit of integration is changed from $-\hbar\omega$ to 0 as DOS ($\rho(E)$) is zero for negative energies. By implementing this change in the second term in parenthesis in Equations (3.9) and (3.22), and by factoring out the I_{com}, the variance of the voltage takes the form

$$\langle V^2(t) \rangle = \int_0^{\infty} d\omega \frac{\hbar\omega}{\pi} R(\omega) \frac{e^{\beta\hbar\omega} + 1}{e^{\beta\hbar\omega} - 1}. \tag{3.25}$$

It can be further simplified as

$$\langle V^2 \rangle = \frac{2}{\pi} \int_0^\infty d\omega R(\omega) E(\omega, T), \tag{3.26}$$

with $E(\omega, T) = \hbar\omega/2 + \hbar\omega/(e^{\hbar\omega/k_B T} - 1)$. This is known as the Callen-Welton theorem [3]. In the high temperature limit, $k_B T \gg \hbar\omega$, $E(\omega, T)$ can be approximated by $k_B T$. This leads to $\langle V^2 \rangle = \frac{2}{\pi} k_B T \int_0^\infty d\omega R(\omega)$ which is nothing but the Nyquist result, but now with the correct coefficient! Callen-Welton's theorem gives the quantum mechanical justification of the Nyquist theorem. The important point that is to be recognized is that the equilibrium voltage fluctuations has "hidden" information regarding the transport coefficients and vice versa. This is one of the "avatars" of the FDT. A much more general formulation was given by Kubo [4]. Kubo derived the FDT using statistical mechanical arguments. The method is different and, of course, the end result is the same. However, Kubo's methods have much more wider practical applicability, especially his formula which connects transport coefficients with integrals of the time correlation functions.

3.2 Kubo's formalism

Transport coefficients can be neatly expressed in terms of correlation functions. As far as one is restricted to the linear response regime, Kubo's formula is the most widely used one and represents the possibility of calculating transport coefficients in any interacting particle system, formally. In this way it goes beyond the Boltzmann equation approach which requires a weakly interacting system of particles. We follow Kubo's original papers [4] for the treatment below. Only the quantum mechanical treatment is given here and the case of electrical transport is worked out in detail which is our object of study here. At the end of this section, a derivation of FDT (Callen-Welton's result) is given using Kubo's method. In the following we consider the Kubo formalism.

This will also set the stage for our next chapter on Memory Function (MF) formalism.

Consider a system with Hamiltonian H. This system is under the influence of an external perturbing field (say, an electric field is applied). Let the perturbation Hamiltonian is given by $H' = -AF(t)$. Where A is an operator to which the external field couples (like electric dipole moment in the case of applied electric field) and $F(t)$ represents an applied time dependent external field (like an AC electric field). Let ρ be the density matrix of the unperturbed system, clearly in equilibrium $[H, \rho] = 0$. Let the density matrix shifts to ρ' under the influence of the applied field. The Liouville-von Neumann equation is given by:

$$\frac{d\rho'(t)}{dt} = \frac{1}{i\hbar}[H + H'(t), \rho'(t)]. \tag{3.27}$$

Let the perturbation be switched on adiabatically in the infinite past at $t = -\infty$. Thus $\rho'(-\infty) = \rho$. As the perturbation is weak one can write $\rho'(t) = \rho + \Delta\rho(t)$ $(||\Delta\rho|| << ||\rho||)$. Then the equation of motion for $\Delta\rho$ is

$$\frac{d\Delta\rho(t)}{dt} = \frac{1}{i\hbar}[H, \Delta\rho(t)] - \frac{1}{i\hbar}F(t)[A, \rho]. \tag{3.28}$$

Here terms retained are linear in $F(t)$. Define an operator $\mathcal{L} = \frac{1}{i\hbar}[H,]$ such that $\mathcal{L}O = \frac{1}{i\hbar}[H, O]$. Multiplying from the left the above equation with an operator integrating factor $OIF = e^{-\mathcal{L}t}$ leads to

$$\frac{d}{dt}(e^{-\mathcal{L}t}\Delta\rho(t)) = -\frac{1}{i\hbar}e^{-\mathcal{L}t}[A, \rho]F(t). \tag{3.29}$$

With the solution

$$\begin{aligned} \Delta\rho(t) &= -\frac{1}{i\hbar}\int_{-\infty}^{t} dt' e^{\mathcal{L}(t-t')}[A, \rho]F(t') \\ &= -\frac{1}{i\hbar}\int_{-\infty}^{t} dt' e^{-i(t-t')H/\hbar}[A, \rho]e^{i(t-t')H/\hbar}F(t')dt'. \end{aligned} \tag{3.30}$$

The last equality in the above equation can be easily derived by observing $e^{\mathcal{L}t}O = (1 + \mathcal{L}t + \frac{1}{2!}\mathcal{L}^2t^2 + ...)O = O + \frac{1}{i\hbar}t[H, O] + \frac{1}{2!}(\frac{1}{i\hbar})^2t^2[H, [H, O]] + ...$

As $\frac{dO}{dt} = \frac{1}{i\hbar}[O, H]$, the last series represents the Taylor expansion of $O(-t)$ around $t = 0$. Thus, the last equality in the above equation follows (remember that the time evolution in the Heisenberg representation is given by $O(t) = e^{iHt/\hbar}Oe^{-iHt/\hbar}$).

Due to this perturbation let an observable B changes by an amount $\Delta B(t)$ evidently

$$\Delta B(t) = Tr(\Delta\rho(t)B) = -\frac{1}{i\hbar}\int_{-\infty}^{t} dt' tr(e^{-i(t-t')H/\hbar}[A,\rho]e^{i(t-t')H/\hbar}B)F(t')dt'. \tag{3.31}$$

By using the cyclic property of trace $tr(ABC) = tr(BCA)$ and writing $B(t - t') = e^{i(t-t')H/\hbar}Be^{-i(t-t')H/\hbar}$ the induced change in the equilibrium operator takes the form:

$$\Delta B(t) = -\frac{1}{i\hbar}\int_{-\infty}^{t} dt' tr([A,\rho]B(t-t'))F(t'). \tag{3.32}$$

If $F(t) = \delta(t)$, i.e., a delta function of unit strength (perturbing field acts for a very short interval of time), then the induced change in the observable has a special name called the response function ϕ for that observable

$$\Delta B(t) = -\frac{1}{i\hbar}tr([A,\rho]B(t)) \equiv \phi_{BA}(t). \tag{3.33}$$

Thus, we have an important relation that we will use frequently

$$\boxed{\phi_{BA}(t) \equiv -\frac{1}{i\hbar}tr([A,\rho]B(t)).} \tag{3.34}$$

In general

$$\Delta B(t) = \int_{-\infty}^{t} dt' \phi_{BA}(t-t')F(t'). \tag{3.35}$$

And if $F(t) = \lim_{\epsilon\to 0^+}\Theta(-t)e^{\epsilon t}$, i.e., a unit step function (perturbing field of the unit strength is applied from infinite past adiabatically and is switched off at $t = 0$), then the induced change in the observable has a special name called the relaxation function Φ defined as

$$\Delta B(t) = \int_{-\infty}^{0} dt' \phi_{BA}(t-t')e^{\epsilon t'} = \boxed{\int_{t}^{\infty} dt' \phi_{BA}(t')e^{-\epsilon t'} \equiv \Phi_{BA}(t).} \tag{3.36}$$

And if $F(t) = \lim_{\epsilon\to 0^+} F_0 \cos(\omega t)e^{\epsilon t}$ (a periodic perturbation adiabatically switched on), then one can write

$$\begin{aligned} \Delta B(t) &= \lim_{\epsilon\to 0^+} \int_{-\infty}^{t} dt'\phi_{BA}(t-t')F_0\cos(\omega t')e^{\epsilon t'} \\ &= \lim_{\epsilon\to 0^+} Re \int_{-\infty}^{t} dt'\phi_{BA}(t-t')F_0 e^{-i\omega t'+\epsilon t'}. \end{aligned} \tag{3.37}$$

Changing integration variables to $\tau = t - t'$ leads to $\Delta B(t) = \lim_{\epsilon\to 0^+} Re(\chi_{BA}(\omega)F_0 e^{-i\omega t+\epsilon t})$ where

$$\boxed{\chi_{BA}(\omega) = \lim_{\epsilon\to 0^+} \int_0^{\infty} dt\phi_{BA}(t)e^{i\omega t-\epsilon t}} \tag{3.38}$$

is called the complex admittance. A very useful equality can be found between complex admittance and relaxation function by integrating the above equation by parts. Let exponential be the first function and ϕ_{BA} be the second. Define $\tilde{\phi}_{BA}(t) = \int \phi_{BA}(t)dt$. By observing $\Phi_{BA}(t) = \tilde{\phi}_{BA}(\infty) - \tilde{\phi}_{BA}(t)$ one can show that

$$\chi_{BA}(\omega) = \Phi_{BA}(0) - i\omega \int_0^{\infty} dt\Phi_{BA}(t)e^{i\omega t}. \tag{3.39}$$

This is very useful equation from experimental perspective, for example, one can measure response function of a system; thus, relaxation function and admittance can be computed from the above equation by numerical integration.

3.2.0.1 Kubo's formula for conductivity

Kubo[4] constructs a very useful representation for the response and relaxation functions. Observe the following identity:

$$e^{\beta H}[A, e^{-\beta H}] = \int_0^{\beta} d\lambda e^{\lambda H}[H, A]e^{-\lambda H}. \tag{3.40}$$

This can be proved by writing the left-hand side as $e^{\beta H}Ae^{-\beta H} - A = A(-i\hbar\beta) - A$ using $A(t) = e^{itH/\hbar}Ae^{-itH/\hbar}$ and the right-hand side as $\int_0^{\beta} d\lambda(HA(-i\hbar\lambda) - A(-i\hbar\lambda)H) = \int_0^{\beta} d\lambda[H, A(-i\hbar\lambda)]$. In the last integrand substitute $t = -i\hbar\lambda$ and use $\dot{A}(t) = \frac{1}{i\hbar}[A, H]$. This leads to $\int_0^{-i\hbar\beta} dt\dot{A}(t) = A(-i\hbar\beta) - A$ which is equal to the left-hand side. QED. Using $\rho = \frac{-\beta H}{Z}$

the above identity can also be written as $[A, \rho] = \int_0^\beta d\lambda \rho e^{\lambda H}[H, A]e^{-\lambda H} = -i\hbar \int_0^\beta d\lambda \dot{A}(-i\hbar\lambda)$. We showed that (Equation (3.33))

$$\phi_{BA}(t) = -\frac{1}{i\hbar} tr([A, \rho]B(t)). \tag{3.41}$$

substituting for $[A, \rho]$ leads to

$$\boxed{\phi_{BA}(t) = \int_0^\beta d\lambda tr(\rho \dot{A}(-i\hbar\lambda)B(t)) = \int_0^\beta d\lambda \langle \dot{A}(-i\hbar\lambda)B(t)\rangle.} \tag{3.42}$$

Here we define ensemble average as $tr(\rho...) = \langle ... \rangle$. The last equation is an extremely useful representation for the response function. Similarly, the relaxation function can also be written in this form. Let us apply this to the case of electrical conduction problem. Let H be the Hamiltonian of a piece of metal, and let the perturbation Hamiltonian be $H' = -\sum_i e\, \mathbf{r_i}.\mathbf{E(t)}$(assume that the metal is under the influence of a time dependent electric field which is position independent). Let the applied field be in the $\nu-$direction and current density $J_\mu = \frac{1}{V}\sum_j e\dot{r}_{j,\mu}(t)$ is measured in the $\mu-$direction:

$$\begin{aligned} \phi_{\mu\nu}(t) &= V \int_0^\beta d\lambda \langle \frac{1}{V}\sum_i e\dot{r}_{i,\nu}(-i\hbar\lambda)\frac{1}{V}\sum_j e\dot{r}_{j,\mu}(t)\rangle \\ &= V \int_0^\beta d\lambda \langle J_\nu(-i\hbar\lambda)J_\mu(t)\rangle. \end{aligned} \tag{3.43}$$

Here V is the volume of the sample. Thus complex AC conductivity in this case takes the form

$$\boxed{\sigma_{\mu\nu}(\omega) = \int_0^\infty dte^{i\omega t}\phi_{\mu\nu}(t) = V\int_0^\infty dte^{i\omega t}\int_0^\beta d\lambda \langle J_\nu(-i\hbar\lambda)J_\mu(t)\rangle.} \tag{3.44}$$

From which DC conductivity can be computed

$$\sigma_{\mu\nu}(0) = V\int_0^\infty dt \int_0^\beta d\lambda \langle J_\nu(-i\hbar\lambda)J_\mu(t)\rangle. \tag{3.45}$$

Another very useful formula can be obtained by integrating the above AC conductivity formula by parts

$$\boxed{\sigma_{\mu\nu}(\omega) = \frac{i}{\omega}\phi_{\mu\nu}(0) + \frac{i}{\omega}\int_0^\infty dt\dot{\phi}_{\mu\nu}(t)e^{i\omega t}}. \tag{3.46}$$

The value of $\phi_{\mu\nu}(0) = \frac{ne^2}{m}\delta_{\mu\nu} = \frac{\omega_p^2}{4\pi}\delta_{\mu\nu}$ where ω_p is the plasma frequency (see next section). The function $\phi_{\mu\nu}(0)$ is the equal time current-current correlation function.

3.2.1 Derivation of Drude's formula of conductivity from Kubo's formula

We have presented two derivations of the classical Drude's formula for conductivity of metals. One is based on the equation of motion for the ensemble averaged momentum with dissipation, and the other one is based on the Boltzmann equation under the relaxation time approximation. Namely, we obtained the following result:

$$Re(\sigma(\omega)) = \frac{ne^2}{m}\frac{\tau}{1+\tau^2\omega^2}. \tag{3.47}$$

Here, we obtain the above equation from Kubo's formula (Equation (3.44)) in the classical approximation ($\hbar \to 0$):

$$Re(\sigma_{xx}(\omega)) = \frac{1}{k_BT}V\int_0^\infty dt\cos(\omega t)\langle J_x(0)J_x(t)\rangle, \tag{3.48}$$

the integration over λ is performed. For simplicity, consider that the electric field is applied in the $x-$direction and induced current is also measured in the $x-$direction. In thermal equilibrium let the correlations decay exponentially. This is related to the Markovian approximation (memoryless friction) as we will see in the following:

$$\langle J_x(0)J_x(t)\rangle = \langle |J_x(0)|^2\rangle e^{-t/\tau} \tag{3.49}$$

As we will see, the relaxation time τ is equal to the Drude relaxation time. In thermal equilibrium the spontaneous current generated at one point decays to zero in the time scale of τ. Now $J_x(0) = nev_x$ (where v_x is the velocity of a single electron participating in the flow) and due to thermal equilibrium one has equipartition law $\frac{1}{2}mv^2 = \frac{3}{2}k_BT$ which yields $v_x = \sqrt{\frac{k_BT}{m}}$ (use $v^2 =$

$v_x^2 + v_y^2 + v_z^2$ and the isotropy of space ($v_x = v_y = v_z$)). With this information

$$\langle |J_x(0)|^2 \rangle e^{-t/\tau} = \frac{tr(\rho |J_x(0)|^2 e^{-t/\tau})}{tr(\rho)} = \frac{n^2 e^2 k_B T}{mN} e^{-t/\tau}. \quad (3.50)$$

Here N is the total number of electrons in volume V. Inserting the above equation in Equation (3.48) and with simple algebra (refer to problem no. (7)) one can show that

$$Re(\sigma(\omega)) = \frac{ne^2}{m} \frac{\tau}{1 + \tau^2 \omega^2}. \quad (3.51)$$

Where τ (current relaxation time) is identified with Drude scattering rate. Thus, under the Markovian approximation Kubo's formula leads to Drude's expression for AC conductivity. DC conductivity is obtained by taking $\lim \omega \to 0$. An important piece of information used in the above calculation is the exponential decay of the current-current correlation function. We will investigate it after introducing the generalized Langevin equation at the end of this chapter. That will lead to a deeper understanding of the above.

3.2.2 Fluctuation-Dissipation Theorem (FDT) from Kubo's formalism

The works of Nyquist and Callen-Welton related to FDT were reformulated by Kubo in an elegant mathematical language [4]. As discussed before, the FDT relates the transport coefficients with corresponding correlation functions. It is illustrated here using the particular example of conductivity and current-current correlation function, but the result has wider applicability. It is straightforward to define the classical correlation function between any two observables $A(t_1)$ and $B(t_2)$ as $\langle A(t_1)B(t_2)\rangle$. In quantum mechanics due to the (non)commutability of operators one has to define correlation functions in a symmetrical way:

$$\Psi_{BA}(t_1 - t_2) = \langle \{A(t_1)B(t_2)\} \rangle = \frac{1}{2} \langle A(t_1)B(t_2) + B(t_2)A(t_1) \rangle. \quad (3.52)$$

Fluctuation-Dissipation Theorem (FDT) deals with the connection between complex admittance and the symmetrized (as above) correlation function of the corresponding observables. Kubo establishes FDT as follows. First a connection between response function ϕ and correlation function Ψ is established. To achieve this, define Fourier transforms of these functions as follows:

$$f_{\mu\nu}(\omega) = \int_{-\infty}^{+\infty} dt\phi_{\mu\nu}(t)e^{i\omega t}, \quad and \quad g_{\mu\nu}(\omega) = \int_{-\infty}^{+\infty} dt\Psi_{\mu\nu}(t)e^{i\omega t}. \tag{3.53}$$

Using Kubo's representation of the response function (Equation (3.43)) one can write

$$f_{\mu\nu}(\omega) = \int_{-\infty}^{+\infty} dte^{i\omega t} \int_0^{\beta} d\lambda \langle J_\nu(-i\hbar\lambda)J_\mu(t)\rangle. \tag{3.54}$$

Now, the expression in the ensamble average can be rewritten as $\langle e^{\lambda H}J_\nu(0)e^{-\lambda H}J_\mu(t)\rangle = \langle J_\nu(0)e^{-\lambda H}J_\mu(t)e^{\lambda H}\rangle$ (as the average involves trace operation and cyclic property of the trace can be used). The last expression is nothing but $\langle J_\nu(0)J_\mu(t+i\hbar\lambda)\rangle$. Thus, the above equation is

$$f_{\mu\nu}(\omega) = \int_{-\infty}^{+\infty} dt \int_0^{\beta} d\lambda \langle J_\nu(0)J_\mu(t+i\hbar\lambda)\rangle e^{i\omega(t+i\hbar\lambda)+\lambda\hbar\omega}. \tag{3.55}$$

Shift the integrating variable t to $t' = t + i\hbar\lambda$:

$$f_{\mu\nu}(\omega) = \int_0^{\beta} d\lambda e^{\lambda\hbar\omega} \int_{-\infty+i\hbar\omega}^{+\infty+i\hbar\omega} dt' \langle J_\nu(0)J_\mu(t')\rangle. \tag{3.56}$$

If the mean value of the current-current correlation has no poles (or is an analytical function) in the strip region (of width $\hbar\omega$) in the upper half plane, then the integration limits can be simply written as $-\infty$ to $+\infty$.[‖] The integral wrt λ can be performed

$$f_{\mu\nu}(\omega) = \frac{e^{\beta\hbar\omega}-1}{\hbar\omega} \int_{-\infty}^{+\infty} dt\langle J_\nu(0)J_\mu(t)\rangle. \tag{3.57}$$

Again restart with Equation (3.54) and consider its integrand (the ensemble averaged part), and observe the following set of transformations:

$$\begin{aligned}
\langle J_\nu(-i\hbar\lambda)J_\mu(t)\rangle &= tr(e^{-\beta H+\lambda H}J_\nu(0)e^{-\lambda H}J_\mu(t)) = tr(J_\nu(0)e^{-\lambda H}J_\mu(t)e^{(\lambda-\beta)H}) \\
&= tr(e^{-\lambda H}J_\mu(t)e^{(\lambda-\beta)H}J_\nu(0)) \\
&= tr(e^{-\beta H}e^{-(\lambda-\beta)H}J_\mu(t)e^{(\lambda-\beta)H}J_\nu(0)) \\
&= \langle J_\mu(t-i\hbar[\beta-\lambda])J_\nu(0)\rangle. \qquad (3.58)
\end{aligned}$$

‖The contribution from vertical contours at remote distances would vanish.

Here the cyclic property of the trace operation is used. With this, the Fourier transform of the response function can be rewritten as

$$f_{\mu\nu}(\omega) = \int_0^\beta d\lambda \int_{-\infty}^{+\infty} dt' e^{i\omega t'} \langle J_\mu(t' - i\hbar[\beta - \lambda]) J_\nu(0) \rangle. \tag{3.59}$$

Shift the variable of integration t' to $t = t' - i\hbar[\beta - \lambda]$

$$f_{\mu\nu}(\omega) = \int_0^\beta d\lambda \int_{-\infty - i\hbar(\beta-\lambda)}^{+\infty - i\hbar(\beta-\lambda)} dt e^{i\omega(t + i\hbar[\beta-\lambda])} \langle J_\mu(t) J_\nu(0) \rangle. \tag{3.60}$$

Again by invoking the analyticity of the mean value of current-current correlation the above equation can be written as

$$= e^{-\beta\hbar\omega} \int_0^\beta d\lambda e^{\lambda\hbar\omega} \int_{-\infty}^{+\infty} dt e^{i\omega t} \langle J_\mu(t) J_\nu(0) \rangle. \tag{3.61}$$

By performing the integration over λ one obtains

$$f_{\mu\nu}(\omega) = e^{-\beta\hbar\omega} \left(\frac{e^{\beta\hbar\omega} - 1}{\hbar\omega} \right) \int_{-\infty}^{+\infty} dt e^{i\omega t} \langle J_\mu(t) J_\nu(0) \rangle. \tag{3.62}$$

Add Equations (3.57) and (3.62), and by noticing that the integrand of the time integration is nothing but twice the correlation function (ref. Equation (3.52)) the result can be collected as

$$\hbar\omega f_{\mu\nu}(\omega) \left(\frac{e^{\beta\hbar\omega/2} + e^{-\beta\hbar\omega/2}}{e^{\beta\hbar\omega/2} - e^{-\beta\hbar\omega/2}} \right) = 2 g_{\mu\nu}(\omega) \tag{3.63}$$

or

$$\frac{\hbar\omega}{2} \coth(\beta\hbar\omega/2) f_{\mu\nu}(\omega) = g_{\mu\nu}(\omega) \quad \text{or,} \quad f_{\mu\nu}(\omega) = \frac{1}{E(\omega, T)} g_{\mu\nu}(\omega). \tag{3.64}$$

Here $E(\omega, T) = \frac{\hbar\omega}{2} \coth(\beta\hbar\omega/2)$. It is the main component of the proof. Next, we express transport coefficient (conductivity in the present example) in terms of the response function (i.e., we rewrite Equation (3.38))

$$\sigma_{\mu\nu}(\omega) = \lim_{\epsilon \to 0+} \int_0^\infty dt \phi_{\mu\nu}(t) e^{i\omega t - \epsilon t}. \tag{3.65}$$

Insert the inverse Fourier transform of $\phi_{\mu\nu}$ in the above equation and swap the integrals (which is permitted and the whole expression remains convergent)

$$\sigma_{\mu\nu}(\omega) = \lim_{\epsilon \to 0+} \int_{-\infty}^{+\infty} \frac{d\omega'}{2\pi} f_{\mu\nu}(\omega') \left(\int_0^\infty dt e^{i(\omega - \omega')t - \epsilon t} \right). \tag{3.66}$$

Perform the integration wrt t (the one in parenthesis)

$$\sigma_{\mu\nu}(\omega) = \frac{i}{2\pi}\int_{-\infty}^{+\infty} d\omega' \frac{f_{\mu\nu}(\omega')}{\omega - \omega' + i\epsilon}, \tag{3.67}$$

by using the identity $\lim_{\epsilon\to 0}\frac{1}{x+i\epsilon} = P\frac{1}{x} - i\pi\delta(x)$ (P is for the principle value**) the above equation can be simplified to

$$\sigma_{\mu\nu} = \frac{1}{2}\int_{-\infty}^{+\infty} d\omega' f_{\mu\nu}(\omega')\delta(\omega-\omega') + \frac{i}{2\pi}P\int_{-\infty}^{+\infty} d\omega' \frac{f_{\mu\nu}(\omega')}{\omega-\omega'}. \tag{3.68}$$

Thus,

$$\sigma_{\mu\nu} = \frac{1}{2}f_{\mu\nu}(\omega) + \frac{i}{2\pi}P\int_{-\infty}^{+\infty} d\omega' \frac{f_{\mu\nu}(\omega')}{\omega-\omega'}. \tag{3.69}$$

The above expression has all the detailed information including FDT and Karmers-Kronig relations. To extract that information rewrite $f_{\mu\nu}$ in terms of its symmetric and antisymmetric parts (it will also separate the dissipative and non-dissipative parts of the conductivity tensor). Rewrite the above equation as

$$\begin{aligned}\sigma_{\mu\nu}(\omega) = \frac{1}{2}&\left(\underbrace{\frac{f_{\nu\mu}(\omega)+f_{\mu\nu}(\omega)}{2}}_{symmetric} + \underbrace{\frac{f_{\mu\nu}(\omega)-f_{\nu\mu}(\omega)}{2}}_{antisymmetric}\right) + \frac{i}{2\pi}P\int_{-\infty}^{+\infty} d\omega' \frac{1}{\omega-\omega'} \\ &\times\left(\underbrace{\frac{f_{\nu\mu}(\omega')+f_{\mu\nu}(\omega')}{2}}_{symmetric} + \underbrace{\frac{f_{\mu\nu}(\omega')-f_{\nu\mu}(\omega')}{2}}_{antisymmetric}\right).\end{aligned} \tag{3.70}$$

Notice that $f^*_{\mu\nu}(\omega) = f_{\nu\mu}(\omega)$(refer to Appendix 8.1). With this, it can be easily shown that $f^s_{\mu\nu}(\omega)$ (the symmetric part of $f_{\mu\nu}(\omega)$) is a real quantity while $f^a_{\mu\nu}(\omega)$ (the antisymmetric part of $f_{\mu\nu}(\omega)$) is purely imaginary ($f^{a*}_{\mu\nu}(\omega) = -f^a_{\mu\nu}(\omega)$). Keeping these in mind we separate the above equation as follows:

$$\sigma^s_{\mu\nu}(\omega) = \frac{1}{2}f^s_{\mu\nu}(\omega) + \frac{i}{2\pi}\int_{-\infty}^{+\infty} d\omega' \frac{f^s_{\mu\nu}(\omega')}{\omega-\omega'} \tag{3.71}$$

$$\sigma^a_{\mu\nu}(\omega) = \frac{1}{2}f^a_{\mu\nu}(\omega) + \frac{i}{2\pi}\int_{-\infty}^{+\infty} d\omega' \frac{f^a_{\mu\nu}(\omega')}{\omega-\omega'}. \tag{3.72}$$

$^{**}P\int_{-\infty}^{+\infty} d\omega' \frac{f(\omega')}{\omega-\omega'} = \lim_{\delta\to 0}\left(\int_{-\infty}^{\omega-\delta} d\omega' \frac{f(\omega')}{\omega-\omega'} + \int_{\omega+\delta}^{+\infty} d\omega' \frac{f(\omega')}{\omega-\omega'}\right)$

Thus,

$$Re\ \sigma^s_{\mu\nu}(\omega) = \frac{1}{2} f^s_{\mu\nu}(\omega) = \frac{1}{2E(\omega,T)} g^s_{\mu\nu}(\omega), (\text{use}(3.64)) \tag{3.73}$$

$$Im\ \sigma^a_{\mu\nu}(\omega) = \frac{1}{2i} f^a_{\mu\nu}(\omega) = \frac{1}{2iE(\omega,T)} g^a_{\mu\nu}(\omega) \tag{3.74}$$

$$Im\ \sigma^s_{\mu\nu}(\omega) = \frac{1}{\pi}\int_{-\infty}^{+\infty} d\omega' \frac{Re\ \sigma^s_{\mu\nu}(\omega')}{\omega - \omega'} \tag{3.75}$$

$$Re\ \sigma^a_{\mu\nu}(\omega) = -\frac{1}{\pi}\int_{-\infty}^{+\infty} d\omega' \frac{Im\ \sigma^a_{\mu\nu}(\omega')}{\omega - \omega'} \tag{3.76}$$

The first two equations ((3.73) and (3.74)) in the above list pertain to the FDT and the last two are the famous Kramers-Kronig (KK) relations. The last equalities in the first two equations in the above list are obtained by using Equation (3.64) and we have defined $\frac{\hbar\omega}{2}\coth(\beta\hbar\omega/2) = E(\omega,T)$. We have already encountered this function during our analysis of the Callen-Welton's result (refer to Equation (3.26)). Equation (3.73) above apparently differs from the Nyquist-Callen-Welton FDT (Equation (3.26)). In fact they are exactly similar. To show this consider the symmetric part of the last equation in (3.53)

$$g^s_{\mu\nu}(\omega) = \int_{-\infty}^{+\infty} dt \frac{\Psi_{\mu\nu}(t) + \Psi_{\nu\mu}(t)}{2} e^{i\omega t}. \tag{3.77}$$

We have defined $\Psi_{\mu\nu}(t) = V\frac{1}{2}\langle J_\mu(t)J_\nu(0) + J_\nu(0)J_\mu(t)\rangle$.†† By inserting this definition in the above equation and shifting the time argument (using the cyclic property of the trace operation) in $(\langle J_\nu(0)J_\mu(t)\rangle)$ and $(\langle J_\mu(0)J_\nu(t)\rangle)$, say, the above equation can be rewritten as

$$\begin{aligned} &g^s_{\mu\nu}(\omega) = \\ &\frac{1}{2}V\int_{-\infty}^{+\infty} dt\left(\langle J_\mu(t)J_\nu(0)\rangle\left(\frac{e^{i\omega t}+e^{-i\omega t}}{2}\right) + \langle J_\nu(t)J_\mu(0)\rangle\left(\frac{e^{i\omega t}+e^{-i\omega t}}{2}\right)\right) \\ &= \int_{-\infty}^{+\infty} dt\Psi^s_{\mu\nu}(t)\cos(\omega t), \end{aligned} \tag{3.78}$$

††V is the volume. We measure the induced current density due to the applied field that leads to a term in the Hamiltonian, see Equation (3.43).

where $\Psi^s_{\mu\nu}(t) = \frac{1}{2}V(\langle J_\mu(t)J_\nu(0) + J_\nu(t)J_\mu(0)\rangle) = V\langle\{J_\mu(t)J_\nu(0)\}\rangle$. Considering this, Equation (3.73) can be written as

$$Re\ \sigma^s_{\mu\nu}(\omega) = \frac{1}{2E(\omega,T)}\int_{-\infty}^{+\infty} dt\Psi^s_{\mu\nu}(t)\cos(\omega t). \tag{3.79}$$

To obtain the Callen-Welton result (3.26) that deals with the diagonal part of conductivity (or resistivity) let us consider that the field is applied in the $x-$direction and the induced current is also measured in the $x-$direction. In this situation, the above equation reduces to

$$Re\ \sigma_{xx}(\omega) = \frac{V}{2E(\omega,T)}\int_{-\infty}^{+\infty} dt\langle\{J_x(0)J_x(t)\}\rangle\cos(\omega t). \tag{3.80}$$

Also consider that the conductor is cylindrical in shape with cross-sectional area A and length l. The above equation can be made explicit as

$$2E(\omega,T)Re\left[\frac{1}{Z_{xx}(\omega)}\right] = Re\int_{-\infty}^{+\infty} dt\langle\{I_x(0)I_x(t)\}\rangle e^{-i\omega t} \tag{3.81}$$

Here I_x is the total current operator (not the current density (current = current density × area)) and the impedance $Z_{xx}(\omega) = \frac{\rho_{xx}(\omega)L}{A}$ (ρ is the complex resistivity). Insert the Fourier transform $I_x(t) = \frac{1}{2\pi}\int_{-\infty}^{+\infty} d\omega I_x(\omega)e^{-i\omega t}$ of the current operator into the above equation and integrate over the resulting delta function

$$\frac{2E(\omega,T)R_{xx}(\omega)}{|Z_{xx}(\omega)|^2} = \frac{1}{2\pi}Re\int_{-\infty}^{+\infty} d\omega'\langle\{I_x(\omega')I_x(\omega)\}\rangle. \tag{3.82}$$

It has another useful form

$$Re\ Y(\omega) = \frac{1}{4\pi E(\omega,T)}Re\int_{-\infty}^{+\infty} d\omega'\langle\{I_x(\omega')I_x(\omega)\}\rangle. \tag{3.83}$$

Here $Y(\omega)$ is the complex admittance ($Y(\omega) = 1/Z(\omega)$). Further notice from the definition of the Fourier transform $I_x(t) = \frac{1}{2\pi}\int_{-\infty}^{+\infty} d\omega I_x(\omega)e^{-i\omega t}$, the integration over ω' can be rewritten as

$$\frac{2E(\omega,T)R_{xx}(\omega)}{|Z_{xx}(\omega)|^2} = Re\ \langle\{I_x(0)I_x(\omega)\}\rangle. \tag{3.84}$$

$I_x(0)$ is in time domain. Integrate the above equation over ω from 0 to ∞:

$$2\int_0^{\infty} d\omega Re\ Y(\omega)E(\omega,T) = Re\int_0^{\infty} d\omega\langle\{I_x(0)I_x(\omega)\}\rangle. \tag{3.85}$$

The integration over ω can be simplified as

$$\begin{aligned} I_x(0) &= \frac{1}{2\pi}\int_{-\infty}^{+\infty} d\omega I_x(\omega) = \frac{1}{2\pi}\left(\int_{-\infty}^{0} d\omega I_x(\omega) + \int_{0}^{+\infty} d\omega I_x(\omega)\right) \\ &= \frac{1}{2\pi}\left(\int_{0}^{\infty} d\omega I_x(-\omega) + \int_{0}^{+\infty} d\omega I_x(\omega)\right) \\ &= \frac{1}{2\pi}\left(\int_{0}^{\infty} d\omega I_x^*(\omega) + \int_{0}^{+\infty} d\omega I_x(\omega)\right) \\ &= \frac{1}{\pi} Re \int_{0}^{\infty} d\omega I_x(\omega). \end{aligned} \tag{3.86}$$

Using this information we obtain the Nyquist result (Equation (6) in his paper [1])

$$\langle I_x^2 \rangle = \frac{2}{\pi}\int_0^{\infty} d\omega Re\, Y(\omega) E(\omega, T). \tag{3.87}$$

Also the average value of the square of the current $\langle I_x^2 \rangle$ is independent of time (we are discussing the equilibrium fluctuations). Thus, the time argument is not shown. This completes our discussion of the FDT. We again return back to the issue of electronic transport in the next section, but with an apparently different perspective of diffusion.

3.3 The Drude formula from the Einstein relation

We will first derive the Einstein relation and then using a definition of the diffusion coefficient the Drude formula will be derived. This involves the frequently used assumption that the equilibrium current correlation function decays exponentially with time. The exponential decay of the correlation functions will be justified in the following and the conditions requiring that will be worked out.

The Einstein relation can be best worked out with a model of diffusing colloidal particles in a suspension (Figure 3.2). Let $n(y)$ be the density of colloidal particles at a height y from the bottom of the vessel (Figure 3.2).

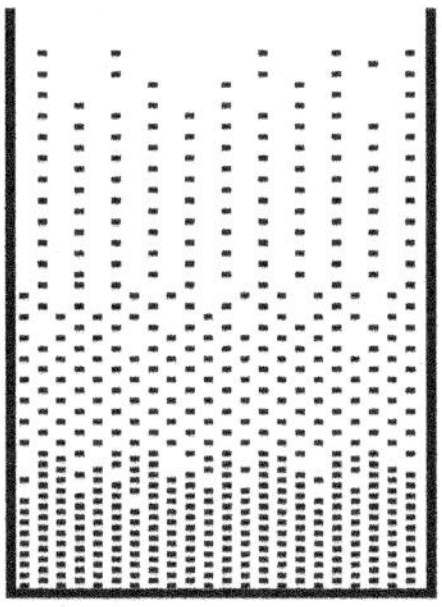

FIGURE 3.2: Diffusing colloidal particles in a suspension.

Assume a uniform density in the horizontal direction at any y. Let m be the mass of a colloidal particle and let $m\gamma$ be the friction (γ is the friction coefficient, i.e., friction per unit mass). Colloidal particles "fall" downward due to gravity with a terminal speed u_t. Using Stokes' law

$$u_t = -\frac{1}{m\gamma}\frac{dU(y)}{dy}. \tag{3.88}$$

Here the gravity force is equal to $-\frac{dU(y)}{dy}$ with $U(y)$ denoting the gravitational potential. Net particle flux in the downward direction is $n(y)u_t$. In equilibrium, this downward flux must be balanced by an equal upward flux. The upward flux is due to the density gradient and is given as $D\frac{dn(y)}{dy}$, where D is the diffusion coefficient. Thus, in equilibrium balance must prevail

$$-n(y)\frac{1}{m\gamma}\frac{dU(y)}{dy} = D\frac{dn(y)}{dy}, \quad \text{or} \quad n(y) = n(0)e^{-U(y)/m\gamma D}, \tag{3.89}$$

where we assumed $U(0) = 0$, and $n(0)$ is the density at the bottom of the vessel. Now, in equilibrium, the distribution $n(y)$ must obey the Boltzmann law (as colloidal particles are in equilibrium with the molecules of the fluid), i.e., $n(y) = n(0)e^{-U(y)/K_BT}$. Thus,

$$D = \frac{k_BT}{m\gamma} = \mu k_BT. \tag{3.90}$$

This relation was derived by Einstein in 1905 and is called the Einstein relation [5]. Here $\mu = \frac{1}{m\gamma}$ is the mobility of a colloidal particle (the ratio

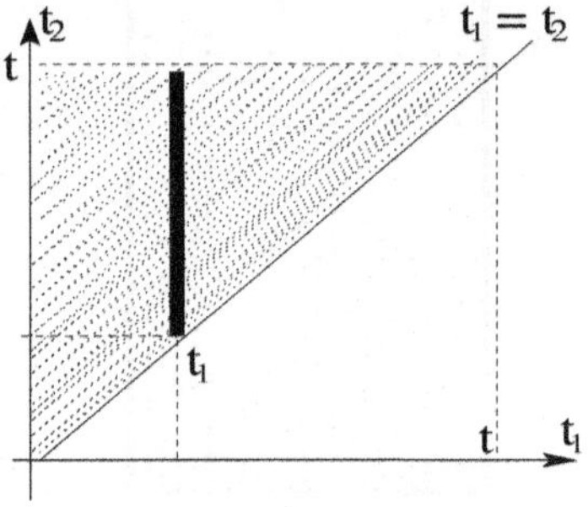

FIGURE 3.3: The volume above the shaded area due to the surface $\langle u(t_1)u(t_2)\rangle$ is half the volume above the total area t^2. The function $\langle u(t_1)u(t_2)\rangle$ is symmetric around the line $t_1 = t_2$.

of terminal speed with impressed force (refer to Equation (3.88))). Now the macroscopic law "diffusion flux= $D\frac{dn(y)}{dy}$" is due to the Brownian motion of the colloidal particles and it can be proved in a very general manner [2] that the distribution for displacement $y(t) - y(0)$ forms a normal distribution with variance proportional to time, i.e., $\langle(y(t) - y(0))^2\rangle = 2Dt$, in the long time limit. Here the long time scale means a time scale much greater than the molecular collision frequency. Thus, $D = \lim_{t\to\infty} \frac{1}{2t}\langle(y(t) - y(0))^2\rangle$. This is a direct consequence of the central limit theorem [2]. As $y(t) - y(0) = \int_0^t dt' u(t')$ the diffusion coefficient can be written as

$$D = \lim_{t\to\infty} \frac{1}{2t} \int_0^t dt_1 \int_0^t dt_2 \langle u(t_1)u(t_2)\rangle. \tag{3.91}$$

Our purpose is to express the mobility μ in terms of the velocity correlation function. If $\langle u(t_1)u(t_2)\rangle$ depends only on $|t_1 - t_2|$, then the following equality can be exploited:

$$\int_0^t dt_1 \int_0^t dt_2 \langle u(t_1)u(t_2)\rangle = 2\int_0^t dt_1 \int_{t_1}^t dt_2 \langle u(t_1)u(t_2)\rangle. \tag{3.92}$$

This can be proved as follows (Figure 3.3). On the Left-Hand Side (LHS) both the integrals have limits from 0 to t. If in the second integral over t_2 the limits are to be varied from t_1 to t instead of 0 to t, then the volume (above the shaded area in Figure 3.3) under the surface $\langle u(t_1)u(t_2)\rangle$[‡‡] will be just half of the volume under the same surface when both the limits are from 0 to t. This is due to the fact that the surface $\langle u(t_1)u(t_2)\rangle$ is symmetric about the line $t_1 = t_2$. Thus, the diffusion coefficient can be written as

$$D = \lim_{t\to\infty} \frac{1}{t} \int_0^t dt_1 \int_0^{t-t_1} dt' \langle u(t_1)u(t_1+t')\rangle, \tag{3.93}$$

where the transformation $t_2 = t_1 + t'$ was made. In the limit $t \to \infty$, the upper limit in the second integral $t - t_1 \to t \to \infty$ and the integrand becomes independent of t_1 (as the correlation function depends only on the difference of the time arguments). The integration over t_1 can be performed yielding

$$D = \int_0^\infty dt \langle u(t_0)u(t_0+t)\rangle, \tag{3.94}$$

t_0 can be any time instant. Now using the Einstein relation $\mu = \frac{1}{m\gamma} = \frac{D}{k_B T}$ the mobility can be expressed as

$$\mu = \frac{1}{k_B T} \int_0^\infty dt \langle u(t_0)u(t_0+t)\rangle. \tag{3.95}$$

This is another formulation of FDT and it is called the first FDT in the literature [2]. It expresses that the transport coefficient (here mobility μ) can be expressed in terms of the velocity-velocity correlation function. Next, we apply this to our special problem of electrical conduction. The terminal speed along the $x-$direction gained by an electron due to impressed electric field is $u_t^x = -\mu(eE_x)$ and the induced current density is $J_x = -neu_t^x = ne^2\mu E$. Thus, electrical conductivity $J_x = \sigma E_x$ is given by $\sigma = ne^2\mu$. From the above expression of the mobility (Equation (3.95)) the electrical conductivity can be expressed as

$$\sigma = ne^2 \frac{1}{k_B T} \int_0^\infty dt \langle u(t_0)u(t_0+t)\rangle. \tag{3.96}$$

This is the classical expression for the conductivity and the Drude formula

---

[‡‡]Imagine the function $\langle u(t_1)u(t_2)\rangle$ is plotted along an axis coming out of the plane of the paper (that is, toward you). It is not difficult to imagine a surface symmetric around $t_1 = t_2$.

can be obtained from it under the assumption of the exponential decay of the correlation functions (Equation (3.49)):

$$\langle u(t_0)u(t_0+t)\rangle = \langle u(t_0)^2\rangle e^{-t/\tau} = \frac{k_B T}{m} e^{-t/\tau}, \tag{3.97}$$

where the equipartition theorem has been used $\frac{1}{2}mu(t_0)^2 = \frac{1}{2}k_B T$ for our 1-D case. Inserting this form of the correlation function into Equation (3.96) yields the well-known Drude formula

$$\sigma = \frac{ne^2\tau}{m}. \tag{3.98}$$

In the present section we proved Equation (3.95) from thermodynamical arguments. However, it is not difficult to see that it can be easily derived from the Kubo formula (Equation (3.44)).[§§] We observe a universal connection that the average time for acceleration in the Drude theory is identical to the relaxation time in the Boltzmann equation under RTA and these are also identical to the time scale over which the current-current correlation function decays.

3.4 The Drude formula from the Langevin equation

In the previous section we introduced the Einstein relation and the Drude formula was obtained under the assumption of exponential decay of the velocity-velocity correlation function. In the present section we introduce another important subject of nonequilibrium statistical mechanics, i.e., the Langevin equation. Our purpose is again the same, we want to write AC conductivity (in a general case) in a form resembling the Drude formula of conductivity. Such representation can be obtained using the memory function formalism which will be our subject in the next chapter. First we will derive the Drude formula from the Langevin equation. Then problems of the

[§§]One would need to replace J_μ by u_x, and J_ν by u_x/V (remember the interaction Hamiltonian in this case), and the conductivity σ by the mobility μ.

Langevin equation will be highlighted that will shed light on the problems of the Drude formula itself. This will lead us to a general formulation using memory function formalism. To understand memory function formalism we have to first consider a phenomenological approach.

The Langevin equation depicts the random motion of a Brownian particle (say a pollen grain in some fluid). The randomness in the motion of a B-particle (Brownian particle) originates from the irregular bombardment of the particle by molecules of fluid. At any temperature, as is well known, fluid molecules are in a continuous random agitation. At one instant of time, more molecules hit the B-particle from one side as compared to the other, and at the other instant, the imbalance topples to the other direction resulting in a random force acting on a B-particle. The origin of the random force can also be viewed from the density fluctuations of fluid at a spatial scale of B-particle size. *Thus, as is well known, the random motion of a B-particle is the direct signature of the incessant molecular randomness.*

In addition to giving a random force to B-particle, molecules of a fluid also provide a drag force to the particle when it tries to dash through the fluid (i.e., inducing drag to a directed motion). The "events" of directed motion originate from the very same imbalance of the forces due to molecules. Thus, the force due to molecules comprises two components: (1) systematic drag force and (2) random force. With this, Newton's second law can be written as

$$m\dot{u}(t) = \underbrace{-m\gamma u(t)}_{systematic\ \ part} + \underbrace{R(t)}_{random\ \ part} . \tag{3.99}$$

This is called the Langevin equation. An important point is that it is not restricted to describe the motion of a Brownian particle, but its regime of applicability is extremely wide (as beautifully expressed by Kubo and presented at the beginning of this chapter).

To apply it to electrons in a metal consider the motion of an electron drifting under the influence of an external AC field. Random motion is induced by impurity potentials and phonons. With external AC field ($E_0 \cos(\omega_0 t)$) the

Langevin equation for an electron takes the form

$$m\dot{u}(t) = -m\gamma u(t) + R(t) + E_0 \cos(\omega_0 t). \tag{3.100}$$

Here m is the mass of an electron and γ is the friction coefficient. Write the above equation in the Laplace domain with $u(z) = \int_0^\infty dt u(t) e^{-st}$

$$-mu(0) + smu(z) = -m\gamma u(s) + R(s) + \frac{E_0}{2}\left(\frac{1}{s - i\omega_0} + \frac{1}{s + i\omega_0}\right). \tag{3.101}$$

or

$$\langle u(s)\rangle = \frac{\langle u(0)\rangle}{s+\gamma} + \frac{E_0}{2m(s+\gamma)}\left(\frac{1}{s - i\omega_0} + \frac{1}{s + i\omega_0}\right). \tag{3.102}$$

The ensemble average of the random force $\langle R(t)\rangle = 0$. The above equation in the long time limit $t >> \frac{1}{\gamma}$ can be written in the form

$$\langle u(t)\rangle = Re\{\mu(\omega_0) E_0 e^{i\omega_0 t}\}, \tag{3.103}$$

where

$$\mu(\omega) = \frac{1}{m}\frac{1}{i\omega + \gamma} \tag{3.104}$$

is called the mobility. We have derived the formula $\sigma = ne^2\mu$ in the previous section. Thus, complex conductivity takes the form

$$\sigma(\omega) = \frac{ne^2}{m}\frac{1}{i\omega + \gamma}. \tag{3.105}$$

The DC conductivity $\sigma(0) = ne^2/m\gamma$. We identify $\frac{1}{\gamma}$ with the Drude scattering rate τ_{Drude}. This is the Drude formula that we derived in Chapters 1 and 2. The question is what new have we gained by introducing the Langevin equation? To understand this, first we have to understand the problems of the Langevin equation itself. The Langevin equation consideration of the Drude formula sheds light on the problems of the Drude formula itself as shown below and further opens the way to a generalization of the Drude formula.

3.5 Problem of the Langevin equation

The solution of the Langevin equation (3.99) can be expressed as

$$u(t) = u(t_0)e^{-\gamma(t-t_0)} + \frac{1}{m}\int_{t_0}^{t} dt' e^{-\gamma(t-t')} R(t'). \tag{3.106}$$

Consider the ensemble average $\langle u(t_0)u(t_0+t)\rangle$, from the above equation it can be easily shown that[¶¶]

$$\langle u(t_0)u(t_0+t)\rangle = \langle u(t_0)^2\rangle e^{-\gamma t} \tag{3.107}$$

where we have used $\langle u(t_0)R(t')\rangle = \langle u(t_0)\rangle\langle R(t')\rangle = 0$ as the latter average is zero. Now consider the time derivative of the above correlation

$$\frac{d}{dt}\langle u(t_0)u(t_0+t)\rangle = -\gamma\langle u(t_0)^2\rangle e^{-\gamma t}. \tag{3.108}$$

In the limit $t \to 0$ the equation can be written as

$$\langle u(t_0)\dot{u}(t_0)\rangle = -\gamma\langle u(t_0)^2\rangle \neq 0. \tag{3.109}$$

Thus we reach on an important conclusion that, according to the Langevin equation, the time derivative of the velocity-velocity correlation function wrt to the initial time (t_0) is not zero. Thus $\langle u(t_0)u(t_0+t)\rangle$ is not independent from the initial time t_0.

This conclusion stands in direct contradiction to the requirement of stationarity in an equilibrium setting. The "requirement of stationarity" states that

$$\frac{d}{dt_0}\langle u(t_0)u(t_0+t)\rangle = 0, \tag{3.110}$$

i.e., the correlation function must be independent from initial time t_0. It must be a function of the difference of time arguments only. On differentiating the above equation by parts and then taking the limit $t \to 0$ one obtains

$$\langle u(t_0)\dot{u}(t_0)\rangle = 0. \tag{3.111}$$

[¶¶] Multiply Equation (3.106) with $u(t_0)$ and take the ensemble average.

This equation contradicts Equation (3.109). Thus there must be a subtle problem with the standard Langevin equation. *A direct conclusion from it leads to the fact that the Drude formula is bound to be invalid in a general setting.* We will see it more clearly in the memory function formalism in the next chapter.

3.6 The generalized Langevin equation and the memory function (time dependent friction coefficient)

Thus we observe that the standard Langevin equation does not comply with the stationarity condition. To rectify this defect we have to consider the situation from a more realistic point of view. We took the friction as $-m\gamma$ and it always acts irrespective to the history of the particle's trajectory. An important point is that it does not take into account the dynamical nature of the interaction of the Brownian particle with fluid molecules. Thus the Langevin equation has to be modified:

$$m\dot{u}(t) = \underbrace{-m\int_{-\infty}^{t} dt'\gamma(t-t')u(t') +}_{\textit{systematic part with memory}} \underbrace{R(t)}_{\textit{random part}} + \underbrace{E(t)}_{\textit{external drive}} \quad . \tag{3.112}$$

This is called the Generalized Langevin Equation (GLE). Here a time dependent friction coefficient $\gamma(t)$ is introduced. Thus the B-particle "remembers" the effect of past collisions! Generally, $\gamma(t)$ decays with time (or memory "fades" with time). Our aim in this section is to derive an equivalent of the Drude formula using GLE and we want to explore under what conditions that "Generalized Drude Formula" (GDF) reduces to the simple Drude formula. It appears that GLE has come from a vacuum! It is a well-known fact in the nonequilibrium statistical mechanical literature that GLE can be derived from the first principles using Zwanzig-Mori projection operators. Here we will not go into that consideration [2], instead observe the following calculation.

Take the Fourier transform $u(t) = \frac{1}{2\pi}\int_{-\infty}^{+\infty} d\omega u(\omega)e^{i\omega t}$ of the GLE (3.112)

$$\begin{aligned} i\omega\frac{1}{2\pi}\int_{-\infty}^{+\infty} d\omega u(\omega)e^{i\omega t} &= -\int_{-\infty}^{t} dt'\left(\frac{1}{2\pi}\int_{-\infty}^{+\infty}\gamma[\omega]e^{i\omega(t-t')}\right) \\ &\times \left(\frac{1}{2\pi}\int_{-\infty}^{+\infty} d\omega' u(\omega')e^{i\omega' t'}\right) \\ &+ \frac{1}{2\pi}\int_{-\infty}^{+\infty} d\omega(E(\omega)+R(\omega))e^{i\omega t}, \end{aligned} \tag{3.113}$$

where $\gamma[\omega]$ is defined as half-Fourier transform (or Fourier-Laplace transform)

$$\gamma[\omega] = \int_0^{\infty} dt\gamma(t)e^{-i\omega t}. \tag{3.114}$$

There is nothing mystical about Fourier-Laplace transform here. Actually $\gamma(t) = finite$ for $t \geq 0$, and $\gamma(t) = 0$ for $t < 0$ (no memory of future). Due to this fact the Fourier transform $\int_{-\infty}^{+\infty} dt\gamma(t)e^{-i\omega t}$ reduces to the half-Fourier transform. The upper limit of the integral over t' can be extended from t to $+\infty$ as $\gamma(-ve) = 0$ and then the integral over t' gives a delta function: $\frac{1}{2\pi}\int_{-\infty}^{+\infty} dt' e^{i(\omega'-\omega)t'} = \delta(\omega'-\omega)$. This delta function removes the integral over ω' and $u(\omega)$ can be written in the form

$$\langle u(\omega)\rangle = \frac{E(\omega)}{m(i\omega+\gamma[\omega])}. \tag{3.115}$$

Here we have taken the ensemble average and used $\langle R(t)\rangle = 0$. Due to time dependent friction the nature of the random force also changes (the correlation will no longer be a delta function in time). However, $\langle R(t)\rangle = 0$ remains. This is due to the common origin of both the effects. Fluctuations and dissipation are connected with each other as we observe in FDT.

If $E(t) = E_0\cos(\omega_0 t)$, then by inverse transformation*** we get

$$\langle u(t)\rangle = Re\mu(\omega_0)E_0 e^{i\omega_0 t}, \tag{3.116}$$

where

$$\mu(\omega) = \frac{1}{m}\frac{1}{i\omega+\gamma[\omega]}, \tag{3.117}$$

***In the algebra one has to use $\gamma[-\omega] = \gamma[\omega]^*$ which can be easily proved from the definition of $\gamma[\omega]$.

is called the mobility. Using the relation between conductivity and mobility ($\sigma(\omega) = ne^2\mu$) the AC conductivity from the generalized Langevin equation can be written as

$$\boxed{\sigma(\omega) = \frac{ne^2}{m}\frac{1}{i\omega + \gamma[\omega]}.} \tag{3.118}$$

This expression is called the Generalized Drude Formula (GDF) and $\gamma[\omega]$ is called the memory function. In the next chapter we will show that the Kubo formula for the AC conductivity can also be written in this form by using Zwanzig-Mori projection operators as first was shown by Mori [2].

Our next aim of this section is to show under what conditions the GDF reduces to the simple Drude formula. As GLE includes the memory dependent friction, if that memory is neglected we expect to get back the simple Drude formula. Thus, let us consider the memory function to be a delta function in time, i.e., $\gamma(t) = \gamma_D\delta(t + i0)$ (because $\gamma(t)$ is defined only for positive t). With this $\gamma[\omega] = \gamma_D$ (from Equation (3.114)) and the GDF (Equation (3.118)) reduces to

$$\sigma(\omega) = \frac{ne^2}{m}\frac{1}{i\omega + \gamma_D}, \tag{3.119}$$

which is nothing but the simple Drude formula. In a realistic physical system $\gamma[\omega]$ can be a complicated function. We will derive an explicit expression for it for electron-phonon/impurity scattering cases in the next chapter using memory function formalism. An important point to remember is that the simple Drude formula is a consequence of memoryless $\gamma(t)$! [Refer to problems (8) and (9) in Chapter 8.]

3.7 When can one use the exponential decay of the velocity-velocity correlation function?

We have used the exponential decay form ($\langle u(t_0)u(t_0+t)\rangle = \langle u^2\rangle e^{-t/\tau}$) of the velocity-velocity correlation function on two accounts. We used this in deriving Drude's formula from the Kubo formalism in Section 3.2.1, then in Section 3.3, where we derived Drude's formula from Einstein's relation. As we promised, now we are going to investigate under what conditions the exponential form can be used.

It is evident from Equations (3.118) and (3.119) that when the Memory Function (MF) in the GLE has no memory ($\gamma(t) = \gamma_D\delta(t)$), we get back the simple Langevin equation and corresponding simple Drude's formula. Thus we want to prove that if the mobility $\mu(\omega)$ from GLE takes the form $\frac{1}{m}\frac{1}{i\omega+\gamma}$, then $\langle u(t_0)u(t_0+t)\rangle$ must be having an exponential decay. To prove this we need to show that GLE leads to

$$\mu(\omega) = \frac{1}{k_B T}\int_0^{\infty} dt e^{i\omega t}\langle u(t_0)u(t_0+t)\rangle. \tag{3.120}$$

Proof: Let us start with the GLE (Equation (3.112)) when there is no external drive and with the initial condition: at an initial time $t = t_0$ initial velocity is $u(t_0)$.

$$\dot{u}(t) = -\int_{t_0}^{t} dt'\gamma(t-t')u(t') + \frac{1}{m}R(t). \tag{3.121}$$

Multiply the above with $u(t_0)$ and take the ensemble average:

$$\langle u(t_0)\dot{u}(t)\rangle = -\int_{t_0}^{t} dt'\gamma(t-t')\langle u(t_0)u(t')\rangle. \tag{3.122}$$

Remember $\langle u(t_0)R(t)\rangle = 0$. Integrate the above equation as

$$\int_{t_0}^{t} dt' e^{-i\omega t'}\langle u(t_0)\dot{u}(t')\rangle = -\int_{t_0}^{t} dt' e^{-i\omega t'}\int_{t_0}^{t'} dt''\gamma(t'-t'')\langle u(t_0)u(t')\rangle. \tag{3.123}$$

On the LHS of the above equation, perform integration by parts, do the change

$\tau = t' - t_0$ in the resulting integral, and then take the limit $t \to \infty$. The LHS will be

$$-e^{i\omega t_0}(\langle u(t_0)^2\rangle - i\omega \int_0^\infty d\tau e^{-i\omega\tau}\langle u(t_0)u(t_0+\tau)\rangle). \tag{3.124}$$

And the RHS of Equation (3.123) can be simplified as follows. First, the limits of the integration over dt'' can be changed from t_0 to t', to t_0 to t. Why? The reason is that $\gamma(-ve) = 0$ (memory is nonzero only for positive time arguments). Thus, this change will add only zero to RHS. Second, insert $\gamma(t'-t'') = \frac{1}{2\pi}\int_{-\infty}^{+\infty} d\omega'\gamma[\omega']e^{i\omega^p(t'-t'')}$. Doing the similar transformation $\tau = t'' - t_0$ and with straightforward algebraic transformations,††† Equation (3.123) can be simplified to

$$\frac{1}{m}\frac{1}{i\omega + \gamma[\omega]} = \frac{1}{m}\frac{1}{\langle u^2\rangle}\int_0^\infty dt e^{-i\omega t}\langle u(t_0)u(t_0+t)\rangle. \tag{3.125}$$

The LHS is simply the mobility $\mu(\omega)$, and in the RHS, use equipartition theorem, and we get back Equation (3.120). QED.

It is very clear that if the correlation function in Equation (3.120) has an exponential decay $\langle u^2\rangle e^{-t/\tau_{Drude}}$ we get back the simple Drude formula (Equation (3.119)). *Thus, our frequently used assumption of exponential decay of the velocity-velocity correlation function is valid when the system does not retain the memory or when it is Markovian in nature. This assumption is generally true when external driving frequency is much smaller than the internal characteristic frequency of the system (like molecular collision frequency in the Brownian particle case):*

$$\frac{1}{\omega} >> \frac{1}{molecular\ collision - frequency}. \tag{3.126}$$

For more details refer to [7]. Thus, the Drude formula is valid only at low frequencies (as compared to internal characteristic frequency(ies)) and is bound to be invalid at higher frequencies. This will be more clear when we will

†††One has to take the limit $t_0 \to -\infty$ to get a delta function from the integral over t' (remember that the correlation function depends only on the difference of the time arguments).

consider an explicit case of electronic conductivity of metals due to phonons or impurities scattering using the memory function formalism of Wölfle and Götze. There we will show that the simple Drude formula is valid only at low frequencies (the characteristic energy scale there will be the Fermi energy).

In the next chapter we are going to deal with memory function formalism and using that we will derive the generalized Drude formula, the most important formula used extensively in recent investigations in strongly correlated systems.

3.8 Summary points

1. We first introduced Nyquist's thermodynamical arguments to establish the FDT and then an alternative derivation by Callen and Welton was presented which provides a quantum mechanical treatment of FDT. With that, qualitative arguments of Nyquist are verified and given a quantum mechanical basis.

2. Next, Kubo's linear response theory was presented and a very important relationship between a given transport coefficient and the corresponding correlation function was established. The case was illustrated using electrical conductivity. Then under the assumption of exponentially decaying current-current correlation function, the Drude formula of conductivity was derived from the Kubo formula. And, finally, the FDT, in its full generality, was established using the Kubo formalism.

3. Einstein's relation ($D = \mu k_B T$) was established using a colloidal suspension. Then by expressing the diffusion coefficient in terms of the velocity-velocity correlation function, the Drude formula of conductivity was again obtained. It required the assumption of the exponential decay of the velocity-velocity correlation function.

4. To understand the regime of applicability of frequently used assumption of the exponential decay of correlation functions, the Langevin equation was introduced. The Drude formula was obtained by yet another method (from the Langevin equation). We observed that the Langevin equation violated the important condition of stationarity and pointed out that the Drude formula is bound to be invalid in a general setting.

5. Next, a more realistic approach was taken where we introduced memory in the friction coefficient. This leads to the generalized Langevin equation and the generalized Drude formula. We observed that when there is no memory in the friction coefficient GLE leads to the simple Drude formula.
6. From GLE we noticed that the assumption of the exponentially decaying correlation function is equivalent to memoryless friction. This is equivalent to the Markovian approximation.
7. This approximation is justified only at low frequencies and the Drude formula is bound to be invalid at high frequencies. A quantitative criterion, by considering characteristic energy scales in a system, can be established using the Zwanzig-Mori-Götze-Wölfle memory function formalism presented in the next chapter.

Bibliography

[1] H. Nyquist, *Phys. Rev.* **32**, 110 (1928).

[2] M. Toda, R. Kubo, N. Saito, *Statistical Physics II: Nonequilibrium Statistical Mechanics*, Springer (2000); R. Zwanzig, *Nonequilibrium Statistical Mechanics*, Oxford University Press (2001).

[3] H. B. Callen and T. A. Welton, *Phys. Rev.* **83**, 34 (1951).

[4] R. Kubo, *J. Phys. Soc. Jpn.* **12**, 570 (1957); *Rep. Prog. Phys.* **29**, 255 (1966).

[5] A. Einstein, *Investigations on the Theory of the Brownian Movement*, BN Publishing (2011).

[6] H. Mori, *Prog. Theor. Phys.* **33**, 423 (1965).

[7] J. L. Lebowitz and P. Resibois, *Phys. Rev.* **139**, A1101 (1965).

Chapter 4

The Zwanzig-Mori-Götze-Wölfle memory function formalism

...We can always assert that an ensemble is just a model of reality that can be confirmed by experiment. This brings us to the essential mystery of statistical mechanics, whether equilibrium or nonequilibrium–why do such models work in the first place?

Robert Zwanzig [3]

In the previous chapter we studied the Kubo formalism, Langevin equation, problems associated with the Langevin equation, and the Generalized Langevin Equation (GLE). From the problems of the Langevin equation we observed that the Drude formula is bound to be invalid in a general setting. GLE leads to Generalized Drude Formula (GDF) in which the Drude scattering rate (i.e., the friction coefficient) acquires a frequency dependence and has general validity. This frequency dependence is due to memory effects in the friction coefficient and when these memory effects are neglected we get back the simple Drude formula from GDF.

In the present chapter we are going to derive GDF from Memory Function (MF) formalism. Explicit calculations for the MF will be performed. MF formalism provides a powerful computational approach of transport coefficient in general. We will investigate about the AC conductivity (dynamical conductivity) and the frequency regime over which GDF of AC conductivity reduces to the simple Drude formula and how that frequency regime depends on the system parameters. We start with an introduction to MF formalism in the following section.

4.1 The Zwanzig-Mori memory function (MF) formalism

In the previous chapter we derived the GDF using GLE. Treatment somewhat looks phenomenological. In fact, as mentioned before, it is possible to motivate the GLE from the microscopic dynamical equations of motion. This important task was accomplished by Mori and Zwanzig in the 1960s [1, 2, 3]. It turns out that the GDF is grounded in the microscopic theory and is on a sound footing. In the present treatment we will not delve into this interesting topic [3].

In this chapter we will highlight a very nice alternative route due to Mori and show that the GDF can also be derived from the microscopic Kubo formula of AC conductivity. To do so, we will need the Mori-Zwanzig projection operators [1, 2, 3, 4]. Let us start with the Kubo formula (3.44)

$$\sigma_{\mu\nu}(\omega) = V\int_0^\infty dt e^{i\omega t}\int_0^\beta d\lambda \langle J_\nu(-i\hbar\lambda)J_\mu(t)\rangle. \tag{4.1}$$

First, for simplicity, we will perform a classical calculation (i.e., substituting $\hbar = 0$ in the above equation). The extension to the quantum case is straightforward. With $\hbar \to 0$ the above equation reduces to

$$\sigma_{\mu\nu}(\omega) = V\beta\int_0^\infty dt e^{i\omega t}\langle J_\nu(0)J_\mu(t)\rangle. \tag{4.2}$$

In classical mechanics, the time evolution of a function of dynamical variables is given by

$$\frac{\partial}{\partial t}f = -\mathcal{L}f, \tag{4.3}$$

where $\mathcal{L}$ is the Liouville operator, given by $\mathcal{L} = \frac{\partial H}{\partial \mathbf{p}}\cdot\frac{\partial}{\partial \mathbf{q}} - \frac{\partial H}{\partial \mathbf{q}}\cdot\frac{\partial}{\partial \mathbf{p}}$. Here H is the total Hamiltonian. The formal operator solution of the above Liouville equation reads

$$f(t) = e^{i\mathcal{L}t} f(0). \tag{4.4}$$

With this, the Kubo formula can be written as

$$\sigma_{\mu\nu}(\omega) = V\beta \int_0^\infty dt e^{i\omega t} \langle J_\nu(0) e^{i\mathcal{L}t} J_\mu(0) \rangle. \tag{4.5}$$

To ensure the convergence of the integral, shift to complex frequencies ($\omega \to \omega + i\epsilon = z$, where ϵ is a tiny positive quantity). Time integration can be performed

$$\sigma_{\mu\nu}(z) = V\beta \langle J_\nu \frac{i}{z + \mathcal{L}} J_\mu \rangle. \tag{4.6}$$

Before we introduce projection operators, we need to define a convenient notation. Let us define "vectors" $|A\rangle$ in "Hilbert space"* with the following definition of a scalar product:

$$\langle A|B\rangle = \langle A^* B\rangle, \tag{4.7}$$

we have introduced $tr(\rho ...) = \langle ... \rangle$ (the statistical ensemble average). The complex conjugation $*$ is for complex operators. The vectors $|A\rangle$ "live" in the "Hilbert space" and we can apply all the techniques of a vector space.† An important point that we have to keep in mind is that we are only using an *analogy* to the techniques that we use in standard quantum mechanics.

With this notation Equation (4.6) can be written as

$$\sigma_{\mu\nu}(z) = V\beta \left\langle J_\nu \left| \frac{i}{z + \mathcal{L}} \right| J_\mu \right\rangle. \tag{4.8}$$

Now we are in a position to introduce our projection operators. Define

$$\mathcal{P} \equiv \sum_{\mu'} \frac{1}{\langle J_{\mu'}|J_{\mu'}\rangle} |J_{\mu'}\rangle\langle J_{\mu'}| \tag{4.9}$$

---

*Here we are only using these concepts as an *analogy* to the techniques that we use in quantum mechanics.

†In this connection, we quote Dieter Forster, "A mathematician would, in addition, question whether this space is separable and complete; a physicist will hope that it is, for all practical purposes" [[4], p. 98]. Our treatment closely follows his treatment of Section 5.1 in [4].

as a projection operator on "vectors" $|J_{\mu'}\rangle$ (i.e., it projects an arbitrary vector J onto $J_{\mu'}$). It is easy to verify that $\mathcal{PP} = \mathcal{P}$ (the essential property of projection). We have calculated the magnitude $\langle J_\mu | J_\mu \rangle = \langle J_\mu J_\mu \rangle = \frac{1}{\beta V}\chi_0$ in Section 3.2.1. The $\chi_0 = ne^2/m$ is the static susceptibility. Introduce another projection operator $\mathcal{Q}$ such that $\mathcal{P} + \mathcal{Q} = \mathbb{1}$. Write $\mathcal{L}$ in Equation (4.8) as $\mathcal{L} = \mathcal{L}(\mathcal{P} + \mathcal{Q})$

$$\sigma_{\mu\nu}(z) = V\beta \left\langle J_\nu \left| \frac{i}{z + \underbrace{\mathcal{LQ}}_{X} + \underbrace{\mathcal{LP}}_{Y}} \right| J_\mu \right\rangle. \tag{4.10}$$

By using the operator identity

$$\frac{1}{X+Y} = \frac{1}{X} - \frac{1}{X} Y \frac{1}{X+Y}, \tag{4.11}$$

$\sigma_{\mu\nu}(z)$ can be separated into two terms

$$\sigma_{\mu\nu}(z) = \underbrace{iV\beta \left\langle J_\nu \left| \frac{1}{z + \mathcal{LQ}} \right| J_\mu \right\rangle}_{1st\ \ term} \underbrace{- iV\beta \left\langle J_\nu \left| \frac{1}{z + \mathcal{LQ}} \mathcal{LP} \frac{1}{z + \mathcal{L}} \right| J_\mu \right\rangle}_{2nd\ \ term}. \tag{4.12}$$

The first term can be expanded $\frac{1}{z(1-\frac{1}{z}\mathcal{LQ})} = \frac{1}{z}(1 + \frac{1}{z}\mathcal{LQ} - \frac{1}{z^2}\mathcal{LQLQ} + ...)$. All the terms in this expansion will be zero except the first. It is easy to see this: $\mathcal{Q}|J_\mu\rangle = \mathbb{1}|J_\mu\rangle - \mathcal{P}|J_\mu\rangle = |J_\mu\rangle - |J_\mu\rangle = 0$. Thus, only the first term will be left, which gives $\frac{\chi_0}{z\beta V}\delta_{\mu\nu}$ (remember, you have to multiply it by $i\beta V$).

For the second term, insert for the "sandwiched" operator $\mathcal{P}$ its expression. The second term will be

$$\beta V \chi_0^{-1} \left\langle J_\nu \left| \frac{1}{z + \mathcal{LQ}} \mathcal{L} \sum_{\mu'} |J_{\mu'}\rangle\langle J_{\mu'}| \frac{1}{z + \mathcal{L}} \right| J_\mu \right\rangle. \tag{4.13}$$

Inserting the expressions for 1st term and 2nd term into Equation (4.12) leads to

$$\boldsymbol{\sigma}(z) = i\chi_0 \frac{1}{z}\mathbb{1} - \frac{1}{z}\boldsymbol{M}(z)\boldsymbol{\sigma}(z), \tag{4.14}$$

or

$$\boxed{\boldsymbol{\sigma}(z) = i\frac{\chi_0}{z + \boldsymbol{M}(z)} = i\frac{\omega_p^2}{4\pi}\frac{1}{z + \boldsymbol{M}(z)}}, \tag{4.15}$$

where we have a defined matrix for conductivity $\boldsymbol{\sigma}$ and a new matrix $\boldsymbol{M}$ called the *the memory matrix*:

$$\boxed{\boldsymbol{M}(z) = V\beta\chi_0^{-1}\left\langle \boldsymbol{J}\left|\frac{z}{z+\mathcal{L}\mathcal{Q}}\mathcal{L}\right|\boldsymbol{J}\right\rangle.} \tag{4.16}$$

We have defined $|\boldsymbol{J}\rangle$ as a column vector. For simplicity, let us assume that an electric field is applied in the x-direction and the induced current is also measured in the same direction. In this case, the above matrix equation reduces to the simple algebraic equation

$$\boxed{\sigma_{xx}(z) = i\frac{\chi_0}{z+M_{xx}(z)}, \quad M_{xx}(z) = V\beta\chi_0^{-1}\left\langle J_x\left|\frac{z}{z+\mathcal{L}\mathcal{Q}}\mathcal{L}\right|J_x\right\rangle.} \tag{4.17}$$

Compare the above equation with the Drude formula (Equation (3.119)) and with the generalized Drude formula from the GLE (Equation (3.118)). Similarity is evident. We changed the notation for memory function from $\gamma[\omega]$ to $M(\omega)$ (as generally used in the literature) but they are the same. In the next section, we will explore under what conditions this form reduces to the simple Drude formula and it will shed light on the importance of the low frequencies in the simple Drude formula. Thus, the complex conductivity from the Kubo formula can be written in the form of GDF. We will also explicitly calculate the MF for impurity scattering and phonon scattering using an ingenious technique due to Götze and Wölfle [5].

The derivation given above is classical ($\hbar \to 0$). It is easy to redo the above calculation by keeping $\hbar$ in Equation (4.1). The derivation is straightforward, one has to change the definition of the scalar product (see the next section or refer to [4]).

4.2 The Götze-Wölfle (GW) formalism

The direct computation of the memory function is an involved task. Götze-Wölfle (GW), in a seminal paper [5], presented a method in which they represented the memory function in terms of the current-current correlation function. Thus, the computation of the memory function reduced to that of the calculation of the current-current correlation function. This is done by treating scattering processes as perturbation. Before presenting their method, we would like to derive the fundamental equation used by them, namely:

$$\boxed{\chi(z) = \chi_0 \frac{M(z)}{z + M(z)}}, \tag{4.18}$$

where $\chi(z)$ is the Fourier-Laplace transform of the current-current correlation function, defined as

$$\chi(z) = iV \int_0^\infty dt e^{izt} \langle [J(t), J(0)] \rangle. \tag{4.19}$$

And $M(z)$ is the MF. Below we will prove the fundamental equation (4.18) using a representation of conductivity by Kubo. Let us define

$$\zeta(t) = i\langle [J(t), J(0)] \rangle. \tag{4.20}$$

Using the cyclic property of trace in the ensemble average, $\zeta(t)$ can be written as

$$itr(\rho J(t) J) - itr(\rho J J(t)) = itr(J(t)[J, \rho]). \tag{4.21}$$

Use Kubo's Identity $e^{\beta H}[J, e^{-\beta H}] = \int_0^\beta e^{\lambda H}[H, J] e^{-\lambda H}$, and write it in the following way:

$$[J, \rho] = \int_0^\beta d\lambda \rho [H, J(-i\lambda)]. \tag{4.22}$$

We set $\hbar = 1$ for simplicity (at any later stage of the calculation, appropriate powers of $\hbar$ can be inserted using dimensional analysis). Using $\dot{A} = i[H, A]$, the above commutator can be written as

$$[J, \rho] = -i \int_0^\beta d\lambda \rho \dot{J}(-i\lambda). \tag{4.23}$$

Inserting this in Equation (4.21), the $\zeta(t)$ takes the form

$$\zeta(t) = \beta\langle \dot{J}(-i\lambda)J(t)\rangle^{\diamond}. \tag{4.24}$$

Here we have introduced a new notation: $\langle ...\rangle^{\diamond} = \frac{1}{\beta}\int_0^{\beta} d\lambda tr(\rho...)$. With this, Equation (4.19) can be written as

$$\chi(z) = V\int_0^{\infty} dte^{izt}\zeta(t) = V\beta\int_0^{\infty} dte^{izt}\langle \dot{J}(-i\lambda)J(t)\rangle^{\diamond}. \tag{4.25}$$

For the next piece of proof, we are going to use an alternative representation of conductivity from Kubo's formula (Equation (3.46)). Inserting the value of $\dot{\phi}$ from Equation (3.43), the conductivity can be written as

$$\sigma(z) = -\frac{\phi(0)}{iz} - \frac{\beta V}{iz}\int_0^{\infty} dte^{izt}\langle J(-i\lambda)\dot{J}(t)\rangle^{\diamond}. \tag{4.26}$$

Our next aim is to find a connection between the last two equations ((4.25) and (4.26)). The only trouble is that the integrands differ with respect to time differentiation. To sort this out, we use the stationarity property of the current-current correlation function, namely, $\langle J(t_0)J(t_0+t)\rangle$ must be independent from the initial time t_0 or

$$\frac{d}{dt_0}\langle J(t_0)J(t_0+t+i\lambda)\rangle^{\diamond} = 0 = \langle \dot{J}(t_0)J(t_0+t+i\lambda)\rangle^{\diamond} + \langle J(t_0)\dot{J}(t_0+t+i\lambda)\rangle^{\diamond}. \tag{4.27}$$

Thus at $t_0 = 0$

$$\langle \dot{J}(0)J(t+i\lambda)\rangle^{\diamond} = -\langle J(0)\dot{J}(t+i\lambda)\rangle^{\diamond}. \tag{4.28}$$

From the above equation it is easy to prove that $\langle \dot{J}(-i\lambda)J(t)\rangle^{\diamond} = -\langle J(-i\lambda)\dot{J}(t)\rangle^{\diamond}$. (Use the cyclic property of the trace operation and notice that the time evolution e^{iH} commutes with the density matrix.) Using this information, Equations (4.25) and (4.26) lead to

$$\sigma(z) = -\frac{\phi(0)}{iz} + \frac{1}{iz}\chi(z). \tag{4.29}$$

We know that (from the section where we derived the Drude formula from the Kubo formalism) $\phi(0) = \beta V\langle J^2\rangle = \frac{ne^2}{m} = \frac{\omega_p^2}{4\pi}$. Thus, the above equation takes the form

$$\sigma(z) = i\frac{\omega_p^2}{4\pi z} - \frac{i}{z}\chi(z). \tag{4.30}$$

We have already expressed the conductivity in terms of the memory function (Equation (4.17)). Again considering (for simplicity) the diagonal matrix element of the conductivity (field applied in x-direction and induced current is also measured in the same direction), the $\chi(z)$ can written as

$$\chi(z) = \frac{\omega_p^2}{4\pi}\frac{M(z)}{z+M(z)} = \chi_0\frac{M(z)}{z+M(z)}. \tag{4.31}$$

This completes our proof of Equation (4.18). Thus, the memory function can be computed if we can compute the Fourier-Laplace transform of the current-current correlation function or $\chi(z)$. *Up to this point, everything is non-perturbative (except the light-matter interaction where we have considered the linear response). The total Hamiltonian of the system still contains two types of contributions:* $H_{total} = H_{electrons} + H_{electron-phonon/impurities}$. *The electron-phonon or electron-impurity interaction causes current degradation and is responsible for electrical resistance. In the next section we discuss a method to deal with these interactions in a perturbative way.*

4.2.1 The Götze-Wölfle (GW) approximation for the memory function

Under weak impurity and/or phonon scattering, GW were able to obtain an approximate representation of the memory function [5]. To obtain this we need Equations-of-Motion (EOM) for the current-current correlation function. Let us introduce, along with GW, the following notation. Consider Equation (4.19) and write it in the following way:

$$\chi_{AB}(z) = \langle\langle A; B\rangle\rangle_z = iV\int_0^\infty dt e^{izt}\langle[A(t), B(0)]\rangle. \tag{4.32}$$

Here A and B are general operators: A is an operator whose linear response is being measured and B is the operator that couples to the external deriving force. We are going to use the following EOMs:

$$z\langle\langle A; B\rangle\rangle_z = -\langle[A, B]\rangle + \langle\langle[A, H]; B\rangle\rangle_z \tag{4.33}$$

$$= -\langle[A, B]\rangle - \langle\langle A; [B, H]\rangle\rangle_z. \tag{4.34}$$

Proof is simple.[‡] In the current case of electrical conduction we are dealing with current-current correlation function. Thus, A and B are current density operators J_1. With this the first EOM reduces to

$$z\chi(z) = \langle\langle C; J_1 \rangle\rangle_z, \tag{4.36}$$

where we have defined a new operator $C = [J_1, H]$ and the equilibrium current-current correlation function $\langle [J_1, J_1] \rangle = 0$. From the second EOM set $A = C$ and $B = J_1$:

$$z\langle\langle C; J_1 \rangle\rangle_z = -\langle [C, J_1] \rangle - \langle\langle C; C \rangle\rangle_z, \tag{4.37}$$

or

$$\langle\langle C; J_1 \rangle\rangle_z = \frac{-\langle [C, J_1] \rangle - \langle\langle C; C \rangle\rangle_z}{z}. \tag{4.38}$$

Substituting $z = 0$ in Equation (4.37) leads to $\langle [C, J_1] \rangle = \langle\langle C; C \rangle\rangle_0$. Using this information and using Equation (4.36), the above Equation leads to

$$z\chi(z) = \frac{1}{z}(\langle\langle C; C \rangle\rangle_0 - \langle\langle C; C \rangle\rangle_z). \tag{4.39}$$

Exact computation of the above correlators ($\langle\langle C; C \rangle\rangle_z$) is not possible, as we will observe in the following that creation and annihilation operators in $\langle\langle C; C \rangle\rangle_z$ are in the Heisenberg representation with the total Hamiltonian in the time evolution operator. It is only under the assumption of weak interaction Hamiltonian (electron-phonon or electron-impurity) that the time evolution in the Heisenberg operator can be represented with the electronic Hamiltonian whose eigen-system is known to us. These lines will be made explicit in the following calculation and there it can be better understood. And under this circumstance, it is also possible to linearize the relation between the memory function and $\chi(z)$. From Equation (4.31), $M(z)$ can be written

---

[‡]We have

$$z\langle\langle A; B \rangle\rangle_z = iV \int_0^\infty dt z e^{izt} \langle [A(t), B] \rangle = V \int_0^\infty dt \langle [A(t), B] \rangle \frac{d}{dt}(e^{izt}). \tag{4.35}$$

Perform integration by parts and use $\dot{A}(t) = i[H, A]$. The result will be required EOM. The second equation can be proved by using the cyclic property of the trace operation.

as

$$M(z) = \frac{z\chi(z)}{\chi_0 - \chi(z)}, \tag{4.40}$$

which can be expanded as $M(z) = \frac{z\chi(z)}{\chi_0(1-\frac{\chi(z)}{\chi_0})} = \frac{z\chi(z)}{\chi_0}(1 + \frac{\chi(z)}{\chi_0} - ....)$. Keeping the leading order term (in view of the above) we obtain an approximate expression for the memory function

$$\boxed{M(z) = \frac{1}{z\chi_0}(\langle\langle C; C\rangle\rangle_0 - \langle\langle C; C\rangle\rangle_z).} \tag{4.41}$$

This is the main equation used by GW for the computation of impurity and phonon scattering (within the free electron approximation, as detailed below).

4.2.2 Impurity scattering

As an application of the above formalism let us consider the case of a free electron gas interacting with dilute randomly distributed impurities. The Hamiltonian is written as

$$\begin{aligned} H_{total} &= H_{electrons} + H_{electron-impurity} \\ H_{electrons} &= \sum_{\mathbf{k}} \epsilon(\mathbf{k}) c^{\dagger}_{\mathbf{k}} c_{\mathbf{k}} \\ H_{electron-impurity} &= \frac{1}{N} \sum_{j,\mathbf{k},\mathbf{k}'} \langle \mathbf{k}|U^j_1|\mathbf{k}'\rangle c^{\dagger}_{\mathbf{k}} c_{\mathbf{k}'}. \end{aligned} \tag{4.42}$$

Here we consider nonmagnetic impurities and electrons do not suffer spin flip during scattering (spin-flip case is a straightforward generalization of this, refer to problem number (10)). N is the number of lattice cells (with cell volume unity) and the sum over j runs over all the randomly distributed impurities with total number N_{imp}. Thus, impurity concentration is $c = \frac{N_{imp}}{N}$. U^j_1 is the scattering potential from the jth impurity.

For the computation of MF we need to compute C and the correlator $\langle\langle C; C\rangle\rangle_z$. The operator C is given by $[J_1, H_{total}]$ (as defined in the previous subsection). The current density J_1 will commute with free electron part of the Hamiltonian (it is a constant of motion in a no scattering case). Current density is given by: $J_1 = \sum_{\mathbf{k}} v_x(\mathbf{k}) c^\dagger_{\mathbf{k}} c_{\mathbf{k}}$, where $v_x(\mathbf{k})$ is the $x-$component of the electron velocity $v_x = \frac{\partial \epsilon(\mathbf{k})}{\partial k_x}$. We consider a case in which an external field is applied in the x-direction and the induced curred is also measured in the same direction. Thus,

$$C = [J_1, H_{electron-impurity}] = \frac{1}{N} \sum_{j,\mathbf{k},\mathbf{k}'} \langle \mathbf{k}|U_1^j|\mathbf{k}'\rangle (v_x(\mathbf{k}) - v_x(\mathbf{k}')) c^\dagger_{\mathbf{k}} c_{\mathbf{k}'}. \quad (4.43)$$

The above calculation is straightforward (use anticommutation properties of $c^\dagger_{\mathbf{k}}$ and $c_{\mathbf{k}}$, and observe for cancellation). Next, we need to evaluate the correlator $\langle\langle C; C\rangle\rangle_z$. Give it a name

$$\phi(z) = \langle\langle C; C\rangle\rangle_z. \quad (4.44)$$

Thus from Equation (4.41)

$$M(z) = \frac{1}{z\chi_0}(\phi(0) - \phi(z)). \quad (4.45)$$

In the computation of $\phi(z)$, one has to deal with the average $\langle\langle c^\dagger_{\mathbf{k}} c_{\mathbf{k}'}; c^\dagger_{\mathbf{p}} c_{\mathbf{p}'}\rangle\rangle$. Expanding the notation

$$\langle\langle c^\dagger_{\mathbf{k}} c_{\mathbf{k}'}; c^\dagger_{\mathbf{p}} c_{\mathbf{p}'}\rangle\rangle = i\int_0^\infty dt e^{izt} \langle [c^\dagger_{\mathbf{k}}(t) c_{\mathbf{k}'}(t), c^\dagger_{\mathbf{p}} c_{\mathbf{p}'}]\rangle. \quad (4.46)$$

Using $c_{\mathbf{k}}(t) = c_{\mathbf{k}} e^{-i\epsilon_{\mathbf{k}} t}$§, the time integration can be easily performed

$$-\frac{1}{z + \epsilon_{\mathbf{k}} - \epsilon_{\mathbf{k}'}} \langle [c^\dagger_{\mathbf{k}} c_{\mathbf{k}'}, c^\dagger_{\mathbf{p}} c_{\mathbf{p}'}]\rangle. \quad (4.47)$$

The thermal ensemble average of the commutator in the above equation can be found (use anticommutation relations for c and $c^\dagger$, and notice that both terms of the commutator lead to a nonvanishing result if $\mathbf{p}' = \mathbf{k}, \quad \mathbf{p} = \mathbf{k}'$). The result will be

$$-\frac{1}{z + \epsilon_{\mathbf{k}} - \epsilon_{\mathbf{k}'}} (f(\mathbf{k}) - f(\mathbf{k}')). \quad (4.48)$$

§As mentioned before, it is the point where the crucial approximation is made. The total Hamiltonian in the time evolution is replaced by $H_{electrons}$ whose eigen-system is known to us. This is possible when the interaction Hamiltonian is small, and we assumed so.

Here $f(\mathbf{k}) = \langle c^\dagger_\mathbf{k} c_\mathbf{k} \rangle$ is the Fermi-Dirac distribution function. Collecting the above information

$$\begin{aligned}\phi(z) &= -2\frac{1}{N^2}\sum_{i,j,\mathbf{k},\mathbf{k}'} \langle \mathbf{k}|U_1^i|\mathbf{k}'\rangle\langle \mathbf{k}'|U_1^j|\mathbf{k}\rangle(v_x(\mathbf{k}) - v_x(\mathbf{k}'))^2 \\ &\times \frac{f(\mathbf{k}) - f(\mathbf{k}')}{z + \epsilon_\mathbf{k} - \epsilon_{\mathbf{k}'}}. \end{aligned} \tag{4.49}$$

The summations run over the total number of impurities N_{imp}. For $i \neq j$, $\phi(z)$ will be proportional to $c^2 = (\frac{N_{imp}}{N})^2$. If the impurity concentration is very low, then this term can be neglected as compared to the term proportional to c (which comes from the case $i = j$). Thus retaining the leading order term

$$\phi(z) = -2\frac{c}{N}\sum_{\mathbf{k},\mathbf{k}'} |\langle \mathbf{k}|U_1|\mathbf{k}'\rangle|^2\{v_x(\mathbf{k}) - v_x(\mathbf{k}')\}^2 \frac{f(\mathbf{k}) - f(\mathbf{k}')}{z + \epsilon_\mathbf{k} - \epsilon_{\mathbf{k}'}}, \tag{4.50}$$

where the summation over the impurity sites i is also performed as there are no correlation effects in the leading order and the summation over the impurity sites just contributes N_{imp} and we have written $c = N_{imp}/N$. Here a factor of 2 appears in the coefficient. This comes from the spin summations (one can consider $\mathbf{k}$ as a joint symbol representing both the wave vector $\mathbf{k}$ and the spin index σ. As there is no spin-flip, the commutator leads to a summation over one spin variable). Further simplification can be done using isotropy in the free electron case $(1/2)m\mathbf{v}^2 = (1/2)m(v_x^2 + v_y^2 + v_z^2) = (3/2)mv_x^2$ and writing $\mathbf{v} = \mathbf{k}/m$

$$\phi(z) = -(2/3)\frac{c}{m^2 N}\sum_{\mathbf{k},\mathbf{k}'} |\langle \mathbf{k}|U_1|\mathbf{k}'\rangle|^2\{\mathbf{k} - \mathbf{k}'\}^2 \frac{f(\mathbf{k}) - f(\mathbf{k}')}{z + \epsilon_\mathbf{k} - \epsilon_{\mathbf{k}'}}. \tag{4.51}$$

From Equation (4.45), $M(z)$ takes the form

$$\begin{aligned}M(z) &= \frac{2}{3z\chi_0}\frac{c}{m^2 N}\sum_{\mathbf{k},\mathbf{k}'} |\langle \mathbf{k}|U_1|\mathbf{k}'\rangle|^2\{\mathbf{k} - \mathbf{k}'\}^2(f(\mathbf{k}) - f(\mathbf{k}')) \\ &\times \left(\frac{1}{z + \epsilon_\mathbf{k} - \epsilon_{\mathbf{k}'}} - \frac{1}{\epsilon_\mathbf{k} - \epsilon_{\mathbf{k}'}}\right). \end{aligned} \tag{4.52}$$

Substitute $z = \omega + i\varepsilon$. On taking the limit $\varepsilon \to 0$ in $z = \omega + i\varepsilon$, the above expression reduces to the imaginary part of the memory function¶

$$ImM(\omega) = \frac{2\pi}{3\omega\chi_0}\frac{c}{m^2 N}\sum_{\mathbf{k},\mathbf{k}'}|\langle\mathbf{k}|U_1|\mathbf{k}'\rangle|^2\{\mathbf{k}-\mathbf{k}'\}^2(f(\mathbf{k})-f(\mathbf{k}'))\delta(\omega+\epsilon_{\mathbf{k}}-\epsilon_{\mathbf{k}'}). \tag{4.53}$$

Consider that the scattering potential is independent of the angle between $\mathbf{k}$ and $\mathbf{k}'$. The above equation can be further simplified in $\omega \to 0$ limit: Change the summations over wave-vectors to integrals over energies by introducing Density of States (DOS), and perform the integrations using the properties of delta functions. The integral over $\epsilon_{\mathbf{k}'}$ can be performed using the properties of the delta function and the other integral over $\epsilon_{\mathbf{k}}$ can be performed using the properties of the delta function resulting from the limit $\omega << \epsilon_F$ in $\frac{f(\epsilon_{\mathbf{k}})-f(\epsilon_{\mathbf{k}}+\omega)}{\omega}$. After some algebra the imaginary part of the MF reads‖

$$ImM(0) = \frac{2}{3}\pi\frac{N_{imp}}{N_e}(U\rho_F)^2\epsilon_F. \tag{4.54}$$

Thus the real part of conductivity from Equation (4.17) is given by

$$Re\sigma(\omega) = \frac{\omega_p^2}{4\pi}\frac{ImM(0)}{\omega^2+(ImM(0))^2}. \tag{4.55}$$

This is nothing but the Drude formula (Section 1.2). We identify that

$$\boxed{ImM(0) = \frac{1}{\tau_{Drude}} = \frac{2}{3}\pi\frac{N_{imp}}{N_e}(U\rho_F)^2\epsilon_F.} \tag{4.56}$$

Thus, the Drude scattering rate increases linearly (in the leading order) with impurity concentration and with the square power of the electron-impurity scattering potential. *The most important point is that we recover the Drude formula in the low frequency limit. The low frequency here means $\omega << \epsilon_F$ (the Fermi energy). If Fermi energy is in eVs then the valid frequency regime is below say infrared frequencies. Thus,*

$$\boxed{\frac{1}{\tau_{Drude}} \text{ is frequency independent, when } \hbar\omega << \epsilon_F.} \tag{4.57}$$

¶Use the identity $\lim_{\varepsilon\to 0}\frac{1}{a+i\varepsilon} = \frac{1}{a} - i\pi\delta(a)$.

‖Details of the calculation are given in the following calculation where U is treated as angle dependent.

Thus the characteristic scale that was mentioned before is the Fermi energy. Next, consider the electron-impurity interaction potential which depends on the wave-vectors and the angle between them. Equation (4.53) reads

$$\begin{aligned} ImM(\omega) &= \frac{2}{3}\pi(\frac{N_{imp}}{N_e})\frac{1}{mN^2}\sum_{\mathbf{k},\mathbf{k}'}|U(\mathbf{k},\mathbf{k}')|^2(\mathbf{k}-\mathbf{k}')^2 \\ &\times \left(\frac{f(\mathbf{k})-f(\mathbf{k}')}{\omega}\right)\delta(\omega+\epsilon_{\mathbf{k}}-\epsilon_{\mathbf{k}'}). \end{aligned} \tag{4.58}$$

Change the summations into integrals (notice that N is the total number of unit cells and each unit cell has a unit volume)

$$\frac{1}{N}\sum_{\mathbf{k}} \longrightarrow \int \frac{d^3k}{(2\pi)^3}. \tag{4.59}$$

With this, the above equation reduces to

$$\begin{aligned} ImM(\omega) &= \frac{2}{3}\pi(\frac{N_{imp}}{N_e})\frac{1}{m}\int\frac{d^3k}{(2\pi)^3}\int\frac{d^3k'}{(2\pi)^3}|U(\mathbf{k},\mathbf{k}',\theta)|^2 \\ &\times (k^2+k'^2-2kk'\cos(\theta))\left(\frac{f(\mathbf{k})-f(\mathbf{k}')}{\omega}\right)\delta(\omega+\epsilon_{\mathbf{k}}-\epsilon_{\mathbf{k}'}). \end{aligned} \tag{4.60}$$

Change to energy using $\epsilon = \frac{k^2}{2m}$ $(\hbar = 1)$ and perform the integration over $\epsilon' = \frac{k'^2}{2m}$ using the properties of the delta function. With simple algebra

$$\begin{aligned} ImM(\omega) &= \frac{2}{3}\pi(\frac{N_{imp}}{Ne})\frac{(2m)^3}{(2\pi)^4}\int_0^\infty d\epsilon\sqrt{\epsilon(\epsilon+\omega)}\int_0^\pi d\theta\sin\theta|U(\epsilon,\epsilon+\omega,\theta)|^2 \\ &\times (2\epsilon+\omega-2\sqrt{\epsilon(\epsilon+\omega)}\cos\theta)\left(\frac{f(\epsilon)-f(\epsilon+\omega)}{\omega}\right). \end{aligned} \tag{4.61}$$

This is a finite frequency result. In the limit $\omega \to 0$, the last factor gives a delta function. The integral over energy thus can be performed with the result

$$\boxed{ImM(0) = \frac{4}{3}\pi(\frac{N_{imp}}{Ne})\frac{(2m)^3}{(2\pi)^4}\epsilon_F^2\int_0^\pi d\theta\sin\theta|U(\epsilon_F,\theta)|^2(1-\cos\theta).} \tag{4.62}$$

It is easy to show that this Equation reduces to Equation (4.56) if one uses the equation for density of states $\rho_F = \frac{4}{(2\pi)^2}\sqrt{2\epsilon_F}m^{3/2}$ and treats the scattering potential as angle independent.

With the angle dependence taken into account, it is clear that the scattering rate ($\frac{1}{\tau} = ImM(0)$) has very tiny contribution from small angle scattering (when θ is small). Thus, small angle scattering is not efficient in the momentum degradation. *The very same conclusion was obtained when we dealt with the solution of the Boltzmann equation beyond Relaxation Time Approximation (RTA) in an isotropic medium where we used an iteration. Thus, the above expression from memory function formalism automatically goes beyond the RTA.*

4.2.3 Phonon scattering

Our second example of a computation of the MF in a concrete situation is the case of electron-phonon scattering in metals. The dominant cause of electrical resistance in metals at ambient temperatures is the electron-phonon scattering. At low temperatures impurity scattering dominates over the phonon scattering. By solving the Bloch-Boltzmann equation we observed that the DC resistivity $\rho \propto T$ when $T >> \Theta_D$, and $\rho \propto T^5$ when $T << \Theta_D$. In the present section we are going to derive these results from MF formalism! But from MF formalism we will get much more. If the sample is very pure, and the temperature is much less than the Debye temperature, then the Holstein mechanism of phonon creation is responsible for the energy absorption at IR frequencies (we have discussed the Holstein mechanism in Chapter 1). We will observe that the Holstein mechanism is a natural outcome of MF formalism.

The Holstein mechanism has been observed experimentally by Joyce and Richards [6], where at IR frequencies an enhanced absorption has been noticed. A qualitative theoretical explanation was given by Falicov in the same paper using the Golden Rule phenomenological approach [6]. However, a more thorough understanding came with Allen's reformulation of the Holstein mechanism using his "Golden Rule" based theories [7]. We will see that the GW formalism can also account for the experimental observations and all the known results related with electron-phonon conductivity obtained using standard Boltzmann equation can be rederived using GW formalism.

To start the calculation of the MF in the present case let us first note the basic assumptions used in the model. Consider that we have a gas of free electrons with spherical Fermi surface interacting with long wavelength longitudinal acoustic phonons (neglect the band structure). The Hamiltonian of the system is given by

$$\begin{aligned} H &= H_0 + H_{el-ph}, \\ H_0 &= H_{el} + H_{ph}, \\ H_{el-ph} &= \sum_{\mathbf{k},\mathbf{k}'} (D(\mathbf{k}-\mathbf{k}')c^\dagger_{\mathbf{k}} c_{\mathbf{k}'} b_{\mathbf{k}-\mathbf{k}'} + h.c.). \end{aligned} \tag{4.63}$$

Here H_0 is the non-interacting part of the Hamiltonian with $H_{el} = \sum_{\mathbf{k}} c^\dagger_{\mathbf{k}} c_{\mathbf{k}}$ and $H_{ph} = \sum_{\mathbf{q}} \omega_{\mathbf{q}} (b^\dagger_{\mathbf{q}} b_{\mathbf{q}} + 1/2)$. The coefficient $D(\mathbf{k}-\mathbf{k}')$ in the electron-phonon Hamiltonian is given by the standard expression for acoustic phonons, i.e., by

$$D(\mathbf{q}) = \frac{1}{\sqrt{2m_{ion} N \omega_{\mathbf{q}}}} q C_q, \tag{4.64}$$

where m_{ion} is the ionic mass, N, as before, is the total number of unit cells, $\omega_{\mathbf{q}}$ is phonon frequency, and C_q is a slowly varying function of q.

Our aim is to compute the MF. Again we set the units with $\hbar = k_B = 1$. We need to compute C and the correlator $\langle\langle C; C\rangle\rangle_z$. (Note that this $C = [J_x, H_{el-ph}]$ is different from the coefficient C_q of the above electron-phonon Hamiltonian. Thus no confusion should arise.) From the definition of J as used before, C can be expressed as

$$C = [J_x, H_{el-ph}] = \sum_{\mathbf{k},\mathbf{k}',\mathbf{p}'} v_x(\mathbf{k}) D(\mathbf{p}-\mathbf{p}') [c^\dagger_{\mathbf{k}} c_{\mathbf{k}}, c^\dagger_{\mathbf{p}} c_{\mathbf{p}'} b_{\mathbf{p}-\mathbf{p}'}] - h.c. \tag{4.65}$$

This can be simplified using the anticommutation properties of $c_{\mathbf{k}}$, and $c^\dagger_{\mathbf{k}}$, and using the commutator identity $[AB, C] = A[B, C] + [A, C]B$ with the result

$$C = \sum_{\mathbf{p},\mathbf{p}'} (v_x(\mathbf{p}) - v_x(\mathbf{p}')) D(\mathbf{p}-\mathbf{p}') c^\dagger_{\mathbf{p}} c_{\mathbf{p}'} b_{\mathbf{p}-\mathbf{p}'} - h.c. \tag{4.66}$$

Our next step is to calculate $\phi(z) = \langle\langle C; C\rangle\rangle_z$. There will be four terms in total. The two diagonal terms (with $b, \quad b$ and $b^\dagger, \quad b^\dagger$) will vanish. The non-vanishing cross terms are

$$\begin{aligned}
\phi(z) &= \\
&- \sum_{\mathbf{k},\mathbf{k}',\mathbf{p},\mathbf{p}'} (v_x(\mathbf{k}) - v_x(\mathbf{k}'))(v_x(\mathbf{p}) - v_x(\mathbf{p}'))D(\mathbf{k}-\mathbf{k}')D^*(\mathbf{p}-\mathbf{p}') \\
&\times \langle\langle c^\dagger_{\mathbf{k}} c_{\mathbf{k}'} b_{\mathbf{k}-\mathbf{k}'}; b^\dagger_{\mathbf{p}-\mathbf{p}'} c^\dagger_{\mathbf{p}'} c_{\mathbf{p}}\rangle\rangle \\
&- \sum_{\mathbf{k},\mathbf{k}',\mathbf{p},\mathbf{p}'} (v_x(\mathbf{k}) - v_x(\mathbf{k}'))(v_x(\mathbf{p}) - v_x(\mathbf{p}'))D^*(\mathbf{k}-\mathbf{k}')D(\mathbf{p}-\mathbf{p}') \\
&\times \langle\langle c^\dagger_{\mathbf{k}'} c_{\mathbf{k}} b^\dagger_{\mathbf{k}-\mathbf{k}'}; b_{\mathbf{p}-\mathbf{p}'} c^\dagger_{\mathbf{p}} c_{\mathbf{p}'}\rangle\rangle.
\end{aligned} \tag{4.67}$$

Consider the correlation function in the first term of the above expression:

$$\langle\langle c^\dagger_{\mathbf{k}} c_{\mathbf{k}'} b_{\mathbf{k}-\mathbf{k}'}; b^\dagger_{\mathbf{p}-\mathbf{p}'} c^\dagger_{\mathbf{p}'} c_{\mathbf{p}}\rangle\rangle = i\int_0^\infty dt e^{izt} \langle [c^\dagger_{\mathbf{k}}(t) c_{\mathbf{k}'}(t') b_{\mathbf{k}-\mathbf{k}'}(t), b^\dagger_{\mathbf{p}-\mathbf{p}'} c^\dagger_{\mathbf{p}'} c_{\mathbf{p}}]\rangle, \tag{4.68}$$

Using $c_{\mathbf{k}}(t) = e^{-i\epsilon_{\mathbf{k}} t}$, the time integration can be done easily, and the commutator can be found using the anticommutation properties of $c^\dagger, \quad c$. After performing the ensemble average, the above expression can be reduced to

$$-\frac{1}{\epsilon_{\mathbf{k}} - \epsilon_{\mathbf{k}'} - \omega_{\mathbf{k}-\mathbf{k}'} + z}\{f_{\mathbf{k}}(1-f_{\mathbf{k}'})(1+n_{\omega_{\mathbf{k}-\mathbf{k}'}}) - f_{\mathbf{k}'}(1-f_{\mathbf{k}})n_{\omega_{\mathbf{k}-\mathbf{k}'}}\}\delta_{\mathbf{p},\mathbf{k}}\delta_{\mathbf{p}',\mathbf{k}'}, \tag{4.69}$$

where, as before, $f_{\mathbf{k}} = \langle c^\dagger_{\mathbf{k}} c_{\mathbf{k}}\rangle$, and $n_{\omega_{\mathbf{k}-\mathbf{k}'}} = \langle b^\dagger_{\mathbf{k}-\mathbf{k}'} b_{\mathbf{k}-\mathbf{k}'}\rangle$. On similar lines the second factor in $\phi(z)$ can be simplified with the net result

$$\begin{aligned}
\phi(z) &= \frac{2}{3}\frac{1}{m^2}\sum_{\mathbf{k},\mathbf{k}'} |D(\mathbf{k}-\mathbf{k}')|^2 |\mathbf{k}-\mathbf{k}'|^2 \{f(1-f')(1+n_-) - (1-f)f'n_-\} \\
&\times \left(\frac{1}{\epsilon - \epsilon' - \omega_- + z} + \frac{1}{\epsilon - \epsilon' - \omega_- - z}\right).
\end{aligned} \tag{4.70}$$

Here a factor of 2 that appears in the coefficient comes from the spin summation. Also, in the above expression we have used shorthand notation to save writing: f means $f(\mathbf{k})$; f' means $f(\mathbf{k}')$; n_- is for $n_{\omega_{\mathbf{k}-\mathbf{k}'}}$, and ω_- is for $\omega_{\mathbf{k}-\mathbf{k}'}$. Then using the definition of the MF (Equation (4.45)), the imaginary part of the memory function can be written as

$$
\begin{aligned}
ImM(\omega) &= \frac{2}{3}\pi\frac{1}{m^2\chi_0}\sum_{\mathbf{k},\mathbf{k}'}|D(\mathbf{k}-\mathbf{k})|^2|\mathbf{k}-\mathbf{k}'|^2(1-f)f'n_- \\
&\times \left(\frac{(e^{\beta\omega}-1)}{\omega}\delta(\epsilon-\epsilon'-\omega_-+\omega) - ...(\text{terms with } \omega\rightarrow-\omega)...\right).
\end{aligned}
\tag{4.71}
$$

In the above, we have used $z=\omega+i\delta$ and have taken the limit $\delta\rightarrow 0$ (use the identity $\lim_{\varepsilon\rightarrow 0}\frac{1}{a+i\varepsilon}=\frac{1}{a}-i\pi\delta(a)$ to obtain the delta functions). To simplify, use the property of the delta function $f(x)\delta(x-a)=f(a)\delta(x-a)$.

Next, insert the expression for $D(\mathbf{k}-\mathbf{k}')$ and convert the sums into integrals (refer to (4.59)). Consider that $\mathbf{k}$ is pointing along the $z-$direction and $\mathbf{k}'$ subtends an angle θ with it. Imagine performing the $\mathbf{k}'$ integral first, and then imagine performing the $\mathbf{k}$ integral in the laboratory frame. Also insert an integral $\int dq\delta(q-|\mathbf{k}-\mathbf{k}'|)$ over q to stratify the calculation:

$$
\begin{aligned}
ImM(\omega) &= \frac{2}{3}\pi\frac{N}{(2\pi)^4 m m_{ion}N_e}\int_0^\infty dq\frac{C_q^2}{\omega_q}q^4\int_0^\infty dkk^2\int_0^\infty dk'k'^2\int_0^\pi d\theta \\
&\times \sin\theta\delta(q-|\mathbf{k}-\mathbf{k}'|)(1-f)f'n_- \\
&\times \left(\frac{(e^{\beta\omega}-1)}{\omega}\delta(\epsilon-\epsilon'-\omega_-+\omega) - ..(\text{terms with } \omega\rightarrow-\omega)..\right).
\end{aligned}
\tag{4.72}
$$

Shift the integration variables to $\epsilon=k^2/2m,\ \ \epsilon'=k'^2/2m$:

$$
\begin{aligned}
& ImM(\omega) = \\
& \frac{2}{3}\pi\frac{Nm^2}{(2\pi)^3 m_{ion}N_e}\int_0^\infty dq\frac{C_q^2}{\omega_q}q^4\int_0^\infty d\epsilon\sqrt{\epsilon}\int_0^\infty d\epsilon'\sqrt{\epsilon'}\int_0^\pi d\theta\sin\theta \\
\times\ & \delta\left(q-\sqrt{2m}\sqrt{\epsilon+\epsilon'-2\sqrt{\epsilon\epsilon'}\cos\theta}\right)(1-f)f'n_- \\
\times\ & \left(\frac{(e^{\beta\omega}-1)}{\omega}\delta(\epsilon-\epsilon'-\omega_-+\omega) - ..(\text{terms with } \omega\rightarrow-\omega)..\right).
\end{aligned}
\tag{4.73}
$$

The integral over θ can be simplified as follows. Due to the presence of the Fermi factors $(1-f)f'$ the integrand has a finite value only in a zone around the Fermi surface (around ϵ_F) and vanishes outside of a strip of width $2k_BT$ (also $\omega << \epsilon_F$). Thus ϵ and ϵ' can be replaced by ϵ_F. With this, the θ integral can be written as

$$\int_0^{\pi} d\theta \sin\theta\delta(q - \sqrt{2}k_F\sqrt{1-\cos\theta}). \tag{4.74}$$

Change to $x = 1 - \cos\theta$ and the integration can be performed with the result $\frac{q}{k_F^2}$ (notice that $0 < q < k_F$, let us say that q (the phonon wavevector) has upper cut-off q_D). Plugging this value $\frac{q}{k_F^2}$ of the θ-integral into the above memory function expression, and then performing the ϵ' integral using the property of the delta functions and performing the integrations over ϵ using elementary methods,** the MF takes the compact form:

$$\begin{aligned} ImM(\omega) &= \frac{1}{8}\pi^3 N(mm_{ion}k_F^5)^{-1}\rho_F^2 \int_0^{q_D} dq q^5 \frac{C_q^2}{\omega_q}\left(\frac{1}{e^{\beta\omega_q}-1}\right) \\ &\times \left\{\frac{(1-\omega_q/\omega)(e^{\beta\omega}-1)}{e^{\beta(\omega-\omega_q)}-1} + ...(\text{terms with } \omega \to -\omega)...\right\}. \end{aligned} \tag{4.75}$$

Thus we rederive GW's equation (54a) in [5] by a slightly different method. For further simplifications we assume (as assumed in their paper [5]) the linear approximation to phonon spectrum $\omega_q = c_s q$ (i.e., the Debye approximation) and C_q is assumed constant $C_q = \frac{1}{\rho_F}$. To extract the information in the above expression it is best to analyze it in special cases. In doing so we will also be able to derive the known results. Consider first the zero frequency limit.

4.2.3.1 DC case

First consider the DC case. In the $\omega \to 0$ limit, the expression in the curly brackets in the above equation reduces to

$$2\beta\frac{\omega_q}{1-e^{-\beta\omega_q}}. \tag{4.76}$$

With this, the MF takes the form

$$ImM(0) = \frac{1}{4}\pi^3 N(mm_{ion}k_F^5)^{-1}\int_0^{q_D} dq \frac{q^5 e^{\beta\omega_q}}{(e^{\beta\omega_q}-1)^2}. \tag{4.77}$$

**Use $\int_{-\infty}^{+\infty} dx \frac{e^x}{e^x+1}\frac{1}{e^{x+a}+1} = \frac{a}{e^a-1}$.

This can be further simplified by writing $x = \beta\omega_q$ and using $\beta\omega_D = \Theta_D = \frac{\Theta_D}{T}$:

$$ImM(0) = \frac{1}{4}\pi^3 N q_D^6 (m m_{ion} k_F^5 \Theta_D)^{-1} (T/\Theta_D)^5 J_5(\Theta_D/T), \tag{4.78}$$

where

$$J_5(y) = \int_0^y dx \frac{x^5 e^x}{(e^x - 1)^2}. \tag{4.79}$$

In the high temperature limit $T >> \Theta_D$, $J_5(y) \simeq \frac{1}{4}y^4$, the imaginary part of the MF reads

$$ImM(0) = \frac{1}{\tau} = \frac{1}{16}\pi^3 N q_D^6 (m m_{ion} k_F^5 \Theta_D)^{-1} (T/\Theta_D) \propto T. \tag{4.80}$$

Thus, in the high temperature approximation (when the temperature is much greater than the Debye temperature), the imaginary part of the memory function or the scattering rate has $T-$linear temperature dependence. This is the same result that was obtained using the Bloch-Boltzmann equation in Chapter 2.

In the low temperature limit $T << \Theta_D$, $J_5(y) \simeq 124.4$, and the MF is

$$ImM(0) = \frac{1}{\tau} = 31.1\pi^3 N q_D^6 (m m_{ion} k_F^5 \Theta_D)^{-1} (T/\Theta_D)^5 \propto T^5. \tag{4.81}$$

This again reproduces the famous Bloch-Boltzmann's T^5 law of phonon induced electrical resistivity $\rho = \frac{1}{\sigma} = \frac{m}{N_e e^2 \tau} \propto T^5$ (refer to the solution of the Bloch-Boltzmann equation in this regime in Chapter 2). Thus, all the known results are reproduced.

4.2.3.2 AC case

A completely general solution is difficult (although the integral over q in the general case can be expressed in terms of PloyLog functions in complicated combinations). We consider the relevant case of high frequencies $\omega >> \Theta_D, T$. In this limit, the expression in curly brackets in Equation (4.75) can be simplified to $e^{\beta\omega_q} + 1$. Then the integral over q can be readily expressed in terms of a special function and the expression for MF takes the form

$$ImM(\omega) \simeq \frac{1}{4}\pi^3 N q_D^6 (m m_{ion} k_F^5)^{-1} (T/\Theta_D)^5 J(\Theta_D/T), \tag{4.82}$$

where

$$J(y) = \frac{1}{2}\int_0^y dx x^4 \coth(x/2). \tag{4.83}$$

In the low temperature limit $T << \Theta_D$, $J(y) \simeq \frac{1}{10}y^5$, the MF takes a constant value

$$ImM(\omega) \simeq \frac{1}{40}\pi^3 N q_D^6 (m m_{ion} k_F^5)^{-1}. \tag{4.84}$$

And in the high temperature limit $T >> \Theta_D$, $J(y) \simeq \frac{1}{4}y^4$. The temperature dependence of MF coincides with the DC limit

$$ImM(\omega) \simeq \frac{1}{16}\pi^3 N q_D^6 (m m_{ion} k_F^5)^{-1}(T/\Theta_D) \propto T. \tag{4.85}$$

The ultra low temperature behavior can be analyzed (at all frequencies) by taking the limit $T \to 0$ in Equation (4.75). The expression in the curly brackets in (4.75) leads to $(1 - \omega_q/\omega)e^{\beta\omega_q}$ if $\omega > \omega_q$ and $(\omega_q/\omega - 1)e^{\beta\omega}$ if $\omega < \omega_q$ and the q integrals can be easily done with the result

$$ImM(\omega) \simeq \frac{1}{240}\pi^3 N q_D^6 (m m_{ion} k_F^5 \Theta_D)^{-1} \times \begin{cases} (\omega/\Theta_D)^5 & : |\omega| < \Theta_D \\ 6 - 5(\omega_D/\omega) & : |\omega| > \Theta_D. \end{cases} \tag{4.86}$$

Thus, we notice that in the low temperature limit there are strong deviations from the Drude result. In the electron-phonon case the scattering rate does depend on frequency (even in the low frequency limit!). As we discussed in the first chapter, Holstein predicted that at finite frequencies and at zero temperature (where thermally excited phonons are not present) there is a finite absorption of the electromagnetic energy due to a special mechanism where phonons are generated along with electron-hole excitations with the absorption of photon quanta. This process is called the Holstein process. We clearly observe a signature of such processes in the above limiting cases. The reader is advised to investigate it further by performing a calculation along the lines of Allen [7], and then the results can be compared for deeper understanding.

4.3 Summary points

1. In the first section it was shown how one can write the Kubo formula of conductivity in the memory function form. The crucial objects are the Zwanzig-Mori projection operators.

2. Then we introduced the Götze-Wölfle formalism. Their fundamental equation (which relates memory function with current-current correlation function) was proved. Equations of motion were developed for the current-current correlation function and using these equations the memory function was expressed in a closed form in terms of the correlation function of the time derivatives of the current operators.

3. By using a perturbative expansion of the memory function with the ratio of dynamical to static correlation function as a perturbation parameter two special cases of interest were considered: the impurity scattering case and the phonon scattering case.

4. In the impurity scattering case we observed that when the frequency is much less than $\epsilon_F/\hbar$ we get back the Drude formula. Also with angle depended scattering potential we obtained the $1 - \cos\theta$ factor in the scattering rate which signifies the importance of large angle scattering in momentum degradation. This is equivalent to the iterative solution of the Boltzmann equation in Chapter 2.

5. Next, we considered the case of electron-phonon scattering. All the known results (the Grüneisen formula) were obtained in the DC limit. In the AC limit we obtained the scattering rate which is able to address the Holstein mechanism (finite absorption of electromagnetic energy even at zero temperature where no thermally excited phonons exist. Absorption happens through the generation of electron-hole pair along with the generation of a phonon).

Bibliography

[1] R. Zwanzig, *Phys. Rev.* **124**, 983 (1961).

[2] H. Mori, *Prog. Theor. Phys.* **33**, 423 (1965); **34**, 399 (1965).

[3] R. Zwanzig, *Nonequilibrium Statistical Mechanics*, Oxford University Press (2001).

[4] Dieter Forster, *Hydrodynamic Fluctuations, Broken Symmetry, and Correlation Functions*, Advanced Book Classic (Westview Press) (1995).

[5] W. Götze and P. Wölfle, *Phys. Rev. B.* **6**, 1226 (1972).

[6] R. R. Joyce and P. L. Richards, *Phys. Rev. Lett.,* **24**, 1007 (1970).

[7] P. B. Allen, *Phys. Rev. B.,* **3**, 305 (1971).

Chapter 5

The Kohn-Luttinger theory: Quantum mechanical basis of the Bloch-Boltzmann equation

Averaging is a different kind of operation since it associates with a stochastic variable a non-stochastic or "sure" number. Alternatively it may be viewed as a projection in the following way. The set of all stochastic variables contains a subset of variables whose probability density is a delta peak. This subset is isomorphic with the sure numbers of the range and may therefore be identified with them. The operation of taking average is then a projection of the total space of stochastic variables onto this subset.

N. G. van Kampen*.

In the last two chapters we worked on the two formulations of the linear response theory, namely the Kubo formalism and the Zwanzig-Mori formalism. Both approaches have the quantum mechanical basis, and the semiclassical assumption that was used to define the distribution function in the Boltzmann equation was not used. And in special cases the Bloch-Boltzmann results were rederived.

In this chapter we would like to revisit the approximations under which the distribution function in the traditional Bloch-Boltzmann equation was

*In *Stochastic Processes in Physics and Chemistry*, Third Edition, North-Holland Personal Library, (2007).

defined. It will be pointed out that the traditional kinetic equations (like the Bloch-Boltzmann equation) have much more serious problems. We are going to present a formalism due to Kohn and Luttinger [1] which justifies the premises on which the Block-Boltzmann equation (or the traditional kinetic theory) is based. This is done in a very simple model. However, once the model is defined, full rigor is maintained throughout. The formalism itself can be used as a computational tool. Thus, the Kohn-Luttinger formalism has two objectives: (1) it provides a quantum mechanical basis to the Bloch-Boltzmann theory, and (2) it provides a mathematical formalism to tackle the transport problems. The objective of this chapter is to deal with the first one.

In the next section, we start with a discussion on the problems of traditional transport theories. Then we will introduce the model considered by Kohn and Luttinger and list the basic assumptions used by them. Once the preliminaries are set we will present the Kohn-Luttinger formulation.

5.1 The assumptions in the traditional kinetic equations

1. In Chapter 2, before introducing the Boltzmann equation, we "justified" the semiclassical model

$$\begin{aligned} \hbar\dot{\mathbf{k}} &= -e\left(\mathbf{E}(\mathbf{r},t) + \frac{1}{c}(\mathbf{v}_n(\mathbf{k}) \times \mathbf{H}(\mathbf{r},t))\right) \\ \dot{\mathbf{r}}(t) &= \mathbf{v}_n(\mathbf{k}) = \frac{1}{\hbar}\frac{\partial \xi_n(\mathbf{k})}{\partial \mathbf{k}}. \end{aligned} \tag{5.1}$$

   The electron waves were replaced by electron wave-packets. The argument was empirically motivated saying that in the presence of scattering, electrons have finite mean free paths but still the momentum remains a good quantum number (refer to the discussion in Chapter 2).

2. Then by defining the coarse-grained phase space (to respect the uncertainty principle) the distribution function $f(\mathbf{r}, \mathbf{k}, t)$ was defined on that

phase space. That is whenever the momentum $\hbar\mathbf{k}$ and position are measured $\mathbf{r}$ their respective measurement uncertainties ($\Delta\hbar\mathbf{k}$ and $\Delta\mathbf{r}$) are assumed to be bigger than the quantum limit $\Delta\hbar\mathbf{k}\Delta\mathbf{r} >> \hbar$.

3. We tacitly assumed that the occupations of various $\mathbf{k}-$states (the diagonal elements of the distribution function) are sufficient to enquire about the dynamics of the system and to find an average value of any given observable.

5.2 The problems with the assumptions

1. Consider the last assumption in the above section (the first two are discussed in Chapter 2). We know from quantum mechanics that to find the expectation value of an observable or its dynamics one needs not only the occupation of various states, i.e., the diagonal elements, but also one needs the off-diagonal elements. In other words, the full density matrix (ρ) is required

$$\langle O \rangle = tr(O\rho). \tag{5.2}$$

   The Bloch-Boltzmann equation deals with the diagonal elements of ρ. The off-diagonal elements are simply neglected and the information contained in them is not taken into account in determining the expectation values. *The answer to this serious problem will be provided by the Kohn-Luttinger formalism.*

2. Alternatively, in the Bloch-Boltzmann theory, one has to assume that some "natural representation" has been found in which the density matrix is diagonal. Thus, the information is not lost, but there are problems with this. It is not always clear what should be the "natural represen-

tation" in a given problem and it turns out that one obtains different answers with different choices [1].

3. There is another serious problem with the Bloch-Boltzmann equation. It is generally assumed that with time evolution the density matrix will remain diagonal. This is based on a very serious assumption called the "Random Phase Assumption (RPA)" [2]. It says that the off-diagonal elements of the density matrix decay continuously due to decoherence caused by random phases between them when statistical ensemble average over randomness is taken. This will be made clear in the following formalism.

The Kohn-Luttinger program is to show how the traditional kinetic equations like the Bloch-Boltzmann emerge when a full quantum approach is adopted. The idea is to start with a simplest possible model but then remain rigorous throughout. Below we start by defining their model and the basic assumptions used.

5.3 The model and assumptions

1. The system considered is a closed cubical box of volume $V = L^3$ (Figure 5.1). Periodic boundary conditions are imposed that allow a steady flow of current through the system. It has a dilute gas of electrons and randomly distributed scattering centers (impurity atoms).

2. An external electric field is impressed on the system, adiabatically. That is, its strength is turned on very slowly from zero to some maximum value.

3. The system contains very dilute (non)interacting electrons and these

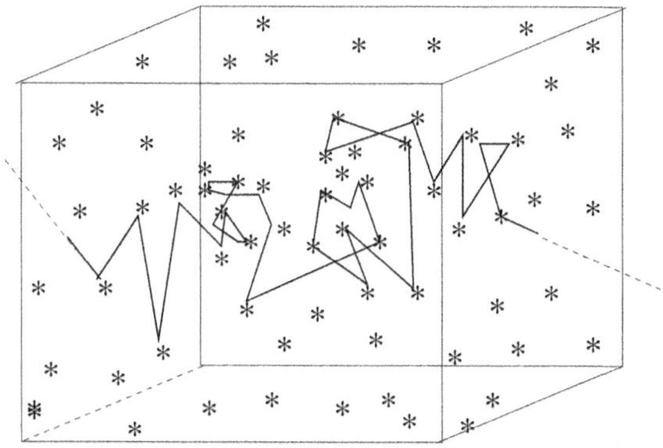

FIGURE 5.1: System is composed of a cubical box of volume V that contains randomly distributed impurities, and it also contains a dilute electron gas. To maintain a steady current flow, periodic boundary conditions are imposed on the faces.

are scattered by "random" rigid impurities. The electron-impurity interaction is characterized by a parameter λ, and the density matrix is developed in power series in λ.

4. The system of electrons is considered to be dilute, so dilute that the quantum statistical effects can be neglected (i.e., the mean occupation of a given state is much less than one and the distribution is not Fermi-Dirac. This is not an important assumption required regarding what is sought at the end of the calculation. Later on this assumption is lifted.)

5.4 The formalism

Since a very dilute gas is considered, the apparent many-body problem reduces to a single-body (one electron) problem. Thus, we write the Hamiltonian for a single electron in the following way:

$$
\begin{aligned}
H_T &= H_0 + H' + H_F & (5.3)\\
H_0 &= p^2/2m \quad \text{(free electron part)} & (5.4)\\
H' &= \sum_{i=1}^{N} \phi(r - r_i) = \lambda \mathscr{V} \quad \text{(el – imp scattering part)} & (5.5)\\
H_F &= -eE_\alpha x_\alpha \quad \text{(electron – field interaction part)}, & (5.6)
\end{aligned}
$$

where r is the position vector for the considered electron.[†] r_i is the position vector for the ith random impurity. The function ϕ denotes the electron-impurity interaction. The net interaction of the considered electron with all the randomly distributed impurities is parametrized with a dimensionless parameter λ which will be our perturbation parameter in the solution of the problem. N is the total number of the "randomly" distributed impurities. The word "random" will be made precise in the following. E_α is the $\alpha = x, y, z$ component of the applied electric field and e is the charge on the electron. When there is no external field $E_\alpha = 0$, define the free field Hamiltonian as $H = H_0 + H'$.

Now consider a collection of N_e electrons each moving under the action of field, i.e., with the Hamiltonian defined above (Equation (5.3)). Let $\psi^i(r,t)$ be the wave-function of an ith electron. Expand this in terms of a complete set of orthonormal basis states $\psi_n(r)$

$$\psi^i(r,t) = \sum_n a_n^i(t)\psi_n(r) \tag{5.7}$$

The total density matrix in the ψ_n representation is given by

$$(\rho_T)_{mn} = \frac{1}{N_e}\sum_{i=1}^{N_e} a_m^i(t)a_n^{i*}(t). \tag{5.8}$$

It can be proved in the following way [3]. Quantum mechanical expectation value with respect to $\psi^i(r,t)$ of any operator $\mathscr{O}(r,t)$ is given by

$$\bar{\mathscr{O}}^i(t) = \int d^3r \psi^{i*}(r,t)\mathscr{O}(r,t)\psi^i(r,t), \tag{5.9}$$

[†]We write r instead of $\mathbf{r}$ to save writing. It is always clear from the context and there should be no confusion.

where "overbar" above the operator $\mathcal{O}$ denotes the quantum mechanical expectation value. This is not enough for a statistical system because the ensemble average of an operator has to be performed (we are dealing with a collection of electrons at some finite temperature T). Thus, we need to perform the ensemble average too

$$\bar{\bar{\mathcal{O}}}(t) = \frac{1}{N_e} \sum_{i=1}^{N_e} \int d^3 r \psi^{i*}(r,t) \mathcal{O}(r,t) \psi^i(r,t), \tag{5.10}$$

where the double bar over the operator $\mathcal{O}$ means two kinds of averages: (1) quantum mechanical expectation and (2) statistical ensemble average [3]. This is what will be measured by some instrument (e.g., current as function of time measured by an ammeter in the above example). Insert the expansion of $\psi^i(r,t)$ in the above mean‡

$$\langle \mathcal{O}(t) \rangle = \sum_{m,n} \underbrace{\left(\frac{1}{N_e} \sum_{i=1}^{N_e} a^i_m(t) a^{i*}_n(t) \right)}_{(\rho_T)_{mn}} \int d^3 r \psi^*_n(r) \mathcal{O}(r,t) \psi_m(r). \tag{5.11}$$

$$\langle \mathcal{O}(t) \rangle = \sum_{m,n} (\rho_T)_{mn} \mathcal{O}_{nm}(t) = \sum_m \langle m | \rho_T \mathcal{O} | m \rangle = tr(\rho_T \mathcal{O}). \tag{5.12}$$

This is the standard procedure of quantum statistical mechanics. *Our objective here is to find out the time evolution equation for the reduced density matrix (reduced density matrix is the density matrix in which linear terms in the applied electric field are retained and ensemble average over the impurities is performed). We will observe that the end equation is the traditional kinetic equation.* To execute this program, let us start with the van Neumann-Liouville equation for the total density matrix

$$i \frac{\partial \rho_T}{\partial t} = [H_T, \rho_T], \qquad (\hbar = 1). \tag{5.13}$$

At $t \to -\infty$, let us assume that there is no external electric field and the system

‡The above mentioned double average will be denoted by a single symbol $\langle ... \rangle$ in what follows.

is in equilibrium at temperature T. The density matrix is the equilibrium density matrix

$$\rho = \frac{1}{Z}e^{-\beta H}, \qquad Z = tr(e^{-\beta H}), \qquad \text{remember,} \quad H = H_0 + H'. \tag{5.14}$$

Then the electric field is very slowly turned on $E_\alpha = E_\alpha^0 e^{st}, \;\; s \to 0^+$, from zero value to the value E_α^0 at $t = 0$. We are interested in the solution of the above equation at $t = 0$. Write the density matrix in the form $\rho_T = \rho + \rho_F$, where ρ_F depends on the applied electric field. In the present treatment ρ_F is restricted to linear order in E_α (for the higher order refer to the appendix of [1]). On substituting $\rho_T = \rho + \rho_F$ into the main equation (5.13), on restricting to the linear order we obtain

$$\frac{\partial \rho_F}{\partial t} = [H_F, \rho] + [H, \rho_F], \tag{5.15}$$

with the initial condition $\rho_F(t \to -\infty) = 0$. Define $H_F = H_1 e^{st}$, with $H_1 = -eE_\alpha^0 x_\alpha$ and insert it in the above equation. Both the above equation and the above-mentioned initial condition are satisfied if we take $\rho_F = fe^{st}$. Here f is the correction to the density matrix at $t = 0$ which is our object of investigation. By inserting $\rho_F = fe^{st}$ in the above equation one obtains

$$isf = [H_1, \rho] + [H, f]. \tag{5.16}$$

To solve it express it in a representation in which H_0 is diagonal. Clearly, the eigenstates of H_0 are plane wave states with periodic boundary conditions

$$\psi_k = \frac{1}{\sqrt{V}}e^{ik.r}, \quad k_\alpha = (2\pi/L)n_\alpha, \quad n_\alpha = 0, \pm1, \pm2, \pm3, ... \tag{5.17}$$

The condition of periodic boundary would look like (Figure 5.2) and leads to the above discretization of k_α values. Eigensystem of H_0 are assumed known $H_0\psi_k = \epsilon_k^0 \psi_k$ with $\epsilon_k^0 = \frac{k^2}{2m}$. In this representation Equation (5.16) takes the form

$$isf_{kk'} = -C_{kk'} + \sum_{k''}(H_{kk''}f_{k''k'} - f_{kk''}H_{k''k'}), \tag{5.18}$$

where the notation $f_{kk'} = \langle k|f|k'\rangle$ for the matrix elements is used, and $C =$

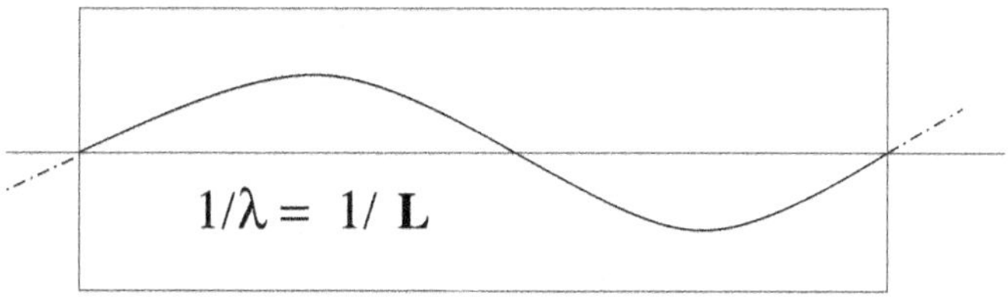

FIGURE 5.2: Periodic boundary conditions to mimic continuity and thermodynamic limit of a system.

$[\rho, H_1]$ is a known matrix as H_1 and ρ are known. On writing $H = H_0 + H'$, and defining $\omega^0_{kk'} = \epsilon^0_k - \epsilon^0_{k'}$ the above equation takes the form

$$(\omega^0_{kk'} - is)f_{kk'} = C_{kk'} + \sum_{k''}(f_{kk''}H'_{k''k'} - H'_{kk''}f_{k''k'}). \tag{5.19}$$

Here the matrix element of the electron-impurity interaction Hamiltonian is given by

$$H'_{kk'} = \frac{1}{V}\int_V d^3re^{-i(k-k').r}\sum_{i=1}^{N}\phi(r-r_i) = \phi_{kk'}\sum_{i=1}^{N}e^{-i(k-k').r_i}, \tag{5.20}$$

where $\phi_{kk'} = \frac{1}{V}\int d^3r'e^{-i(k-k').r'}\phi(r')$. The last equality in the above equation is obtained by shifting the integration variables $r - r_i = r'$. The integral in the definition of $\phi_{kk'}$ is extended from the volume V to all space. This can be done if the range ξ of the potential $\phi(r)$ is assumed to be much smaller than the linear size of the box L, i.e., $\xi << L$ (refer to problem number (11) in the last chapter).

Now the diagonal elements of H' are given by

$$H'_{kk} = N\phi_{kk} \propto N. \tag{5.21}$$

Here N is the number of impurity atoms in volume V. Thus, the diagonal elements of the electron-impurity interaction are proportional to the product of the number of impurity atoms and a constant term coming from the all-space integration of the potential ($\frac{1}{V}\int d^3r\phi(r) = const$). The off-diagonal elements ($k \neq k'$) are

$$H'_{kk'} = \phi_{kk'} \sum_{i=1}^{N} e^{-i(k-k').r_i} \propto \sqrt{N}. \tag{5.22}$$

The proportionality of the off-diagonal elements to $\sqrt{N}$ can be seen as follows. As it was defined, r_i's denote the random positions of the impurity atoms. Due to this randomness, the exponent ($e^{-i(k-k').r_i}$) in the above equation denotes random phases. When these random phases are summed up it leads to a kind of random walk in 3D and it further gives a total sum going as $\sqrt{N}$ (refer to problem number (12) in the last chapter). *This is a very crucial point. Randomness in the positions of the impurities produces a sharp distinction between the diagonal and the off-diagonal elements. The off-diagonal elements are suppressed! Thus, we already see hints why in the traditional transport equations the off-diagonal elements of the density matrix are not taken into account.*

In view of the above observation, split Equation (5.19) into two equations; one dealing with the diagonal elements and the other dealing with the off-diagonal elements. Let us separate the terms in Equation (5.19) with $k'' = k',\ \ k'' = k$ and with $k'' \neq k',\ \ k'' \neq k$, further if $k \neq k'$:

$$\begin{aligned}(\omega^0_{kk'} - is) f_{kk'} &= C_{kk'} + (f_k - f_{k'})H'_{kk'} + f_{kk'}(H'_{k'k'} - H'_{kk}) \\ &+ \sum_{k''(k''\neq k,k')} (f_{kk''}H'_{k''k'} - H'_{kk''}f_{k''k'}).\end{aligned} \tag{5.23}$$

And for $k = k'$ from the above equation we obtain

$$\boxed{-isf_k = C_k + \sum_{k''\neq k} (f_{kk''}H'_{k''k} - H'_{kk''}f_{k''k}).} \tag{5.24}$$

Here we defined $f_k = f_{kk}$ and $C_k = C_{kk}$. As we observed the second term on the RHS of Equation (5.23) is nothing but a constant energy shift term multiplied by the distribution function. Shifting that term to the LHS and redefining $\epsilon_k = \epsilon_k^0 + H'_{kk}$, $\omega_{kk'} = \epsilon_k - \epsilon_{k'}$ we obtain

$$(\omega_{kk'} - is)f_{kk'} = C_{kk'} + (f_k - f_{k'})H'_{kk'} + \sum_{k''(k'' \neq k,k')} (f_{kk''}H'_{k''k'} - H'_{kk''}f_{k''k'}). \tag{5.25}$$

Thus the only effect of the diagonal elements of the perturbation is to shift the energies by a constant term! That can be taken into account by simply redefining the energies. The off-diagonal elements have the important role to play as we will observe in the following. The above equation is the central result and in what follows we would like to find its solution using an iterative procedure in a power series in the interaction strength λ.

First of all we will inspect Equation (5.25) and check how each term in it depends on λ, then leading order terms will be considered (in view of small λ). *As stated before, our aim is to develop a perturbative solution of the equation (5.24)*, this is done through an iterative procedure. Then the leading order result from Equation (5.25) will be substituted into Equation (5.24). Let us first, by inspection, check the dependence of various terms on λ in the above Equation. We start with

$$C_{kk'} = -\langle k|[H_1, \rho]|k'\rangle = -eE_\alpha^0[\rho, x_\alpha]_{kk'}. \tag{5.26}$$

As $\rho = \frac{1}{Z}e^{-\beta H} = \frac{1}{Z}e^{-\beta(H_0+H')} = \frac{1}{Z}e^{-\beta(H_0+\lambda\mathscr{V})}$, it can be expanded in power series in λ. To the leading order (which is zeroth order in λ)

$$C_{kk'} = -e\frac{E_\alpha^0}{Z_0}[e^{-\beta p_\alpha^2/2m}, x_\alpha]_{kk'}. \tag{5.27}$$

The commutator can be very easily evaluated with the result

$$C_{kk'} = ieE_\alpha^0 \left(-\frac{\beta k_\alpha}{m} \underbrace{\frac{1}{Z_0}e^{-\beta\epsilon_k}}_{\rho_k} \right) \delta_{kk'} = ieE_\alpha^0 \left(\frac{\partial \rho_k}{\partial k_\alpha}\right) \delta_{kk'}, \tag{5.28}$$

where $\epsilon_k = \frac{k_\alpha^2}{2m}$. Thus, we notice that the diagonal terms of $C_{kk'}$ begin with the zeroth power in λ, and the off-diagonal terms begin with linear terms in λ. If λ is small the diagonal terms dominate.

Next, we consider the λ dependence of f_k itself. We will see in the later development that $f_k \sim \lambda^{-2}$. A rigorous proof cannot be given here, however, it can be motivated using the Fermi Golden rule in the following way. The transition probability from any initial state k' to a final state k can be written as

$$W_k = \frac{2\pi}{\hbar} \sum_{k'} |\langle k|H'|k'\rangle|^2 \delta(E_k - E_{k'}). \tag{5.29}$$

As $H' \sim \lambda$, this transition probability goes as λ^2. Now, in the linear response regime, if we observe the Bloch-Boltzmann equation (2.61) we immediately see that Φ_k or $g_k \sim \lambda^{-2}$. The quantity that corresponds to g_k here is f_k. Thus, we conclude that $f_k \sim \lambda^{-2}$. With this, one can further observe from Equation (5.25) that the off-diagonal elements $f_{kk'}$ go as λ^{-1}.

Collecting the above information and based on that retaining the leading order terms in Equation (5.25) one obtains

$$f_{kk'} = \frac{(f_k - f_{k'})H'_{kk'}}{\omega_{kk'} - is}. \tag{5.30}$$

Since we are seeking an iterative solution, insert the above equation into Equation (5.24):

$$\boxed{-isf_k = ieE^0_\alpha \frac{\partial \rho_k}{\partial k_\alpha} + \sum_{k'(k'\neq k)} |H'_{kk'}|^2 (f_k - f_{k'}) \left(\frac{1}{\omega_{kk'} - is} - \frac{1}{\omega_{kk'} + is} \right).} \tag{5.31}$$

As will be motivated below, this is the usual transport equation.

5.4.1 Ensemble average over the impurities

The above transport equation is for one member of the ensemble. The positions r_i of the impurity atoms are random. Each member of the ensemble has random distribution of impurities and we need to take the ensemble average. Define

$$\langle \mathscr{O}(r)\rangle_{imp} = \frac{1}{V} \int_V d^3r \mathscr{O}(r), \tag{5.32}$$

as the ensemble average over the impurities (we assume a uniform distribution, and we assume that two or more impurities can take the same position in real space. This is unrealistic but for a point like impurities it is highly unlikely [1]). On performing the ensemble average of the transport equation (5.31) we obtain

$$-isf_k = ieE^0_\alpha \frac{\partial \rho_k}{\partial k_\alpha} + \sum_{k'(k'\neq k)} \langle |H'_{kk'}|^2\rangle_{imp}(f_k - f_{k'}) \left(\frac{1}{\omega_{kk'} - is} - \frac{1}{\omega_{kk'} + is}\right). \tag{5.33}$$

Here we defined the same symbols for f_k and ρ_k, but they represent ensemble averaged quantities (over impurities). We need to find out $\langle |H'_{kk'}|^2\rangle_{imp}$. By using the definition of ensemble average this quantity, by direct calculation, gives

$$\begin{aligned}\langle |H'_{kk'}|^2\rangle_{imp} &= \frac{|\phi_{kk'}|^2}{V^2}\langle \sum_{i,j} e^{-i(r_i - r_j).q}\rangle_{imp} = \frac{|\phi_{kk'}|^2}{V^2}\sum_{i,j}\left(\frac{1}{V}\int d^3r_i\right)^2 \delta_{ij} \\ &= \frac{|\phi_{kk'}|^2}{V^2} N. \end{aligned} \tag{5.34}$$

Here we defined $q = k - k'$. As $k \neq k'$, implies $q \neq 0$. The condition $q \neq 0$ imposes the condition $i = j$, and rest of the calculation is straightforward.

In statistical mechanics a quantity is a well-behaved macroscopic parameter only if fluctuations in it converge in the thermodynamic limit. Let $\mathcal{O}$ be a measurable quantity. If it is a sum function then by the property of statistical independence it can be proved (in a very general setting) that

$$\frac{\sqrt{\langle \mathcal{O}^2\rangle - (\langle \mathcal{O}\rangle)^2}}{\langle \mathcal{O}\rangle} \sim \frac{1}{N}, \tag{5.35}$$

where N is the number of subsystems. Refer to for example [4]. The above equation says that the rms fluctuation goes to zero when the number of constituent goes to infinity (this is at the foundation of statistical thermodynamics, that is why fluctuations are not significant in the thermodynamical systems.[§])

But if we compute the rms fluctuations in $\langle |H'_{kk'}|^2\rangle_{imp}$ we observe that

[§]Pressure exerted by a gas enclosed in a box is constant to a very good approximation, rms fluctuations are less than 10^{-6}.

they are order of one!

$$\frac{\sqrt{\langle(|H'_{kk'}|^2)^2\rangle_{imp} - (\langle|H'_{kk'}|^2\rangle_{imp})^2}}{\langle|H'_{kk'}|^2\rangle_{imp}} \sim 1. \tag{5.36}$$

As can be seen with direct calculation (refer to Appendix 8.2). Thus there seems to be a problem with the ensemble average. However, if a coarse-grained picture is adopted, this problem can be resolved [1]. (Refer to problem (13) in Chapter 8.)

Thus from Equations (5.33) and (5.34) we obtain

$$-isf_k = ieE^0_\alpha \frac{\partial \rho_k}{\partial k_\alpha} + \sum_{k'(k'\neq k)} \left(N\frac{|\phi_{kk'}|^2}{V^2}\right)(f_k - f_{k'})\left(\frac{1}{\omega_{kk'} - is} - \frac{1}{\omega_{kk'} + is}\right). \tag{5.37}$$

If the rate at which the field is turned on (i.e., s) is very small as compared with both the inverse of the relaxation time (i.e., $\frac{1}{\tau_{relax}}$) and with the characteristic inverse electronic time scale $\frac{1}{\tau_a} = \frac{E_F}{\hbar}$, then the LHS in the above equation can be dropped. Thus, when

$$s << \frac{1}{\tau_{relax}} \quad \text{and} \quad \frac{1}{\tau_a} \tag{5.38}$$

one has

$$-ieE^0_\alpha \frac{\partial \rho_k}{\partial k_\alpha} = \sum_{k'(k'\neq k)} \left(N\frac{|\phi_{kk'}|^2}{V^2}\right)(f_k - f_{k'})\left(\frac{1}{\omega_{kk'} - is} - \frac{1}{\omega_{kk'} + is}\right). \tag{5.39}$$

Further if ΔE is the energy spacing in the quasi-continuous spectrum of electronic energy levels and if $s >> \frac{\Delta E}{\hbar}$, then the spectrum can be considered continuous and the summation over k' can be replaced with integration

$$\sum_{k'} = \frac{V}{(2\pi)^3}\int d^3k'. \tag{5.40}$$

On implementing these changes the transport equation takes the form

$$-ieE^0_\alpha \frac{\partial \rho_k}{\partial k_\alpha} = 2\pi \frac{V}{(2\pi)^3}\int d^3k' \left(N\frac{|\phi_{kk'}|^2}{V^2}\right)(f_k - f_{k'})\delta(\omega_{kk'}), \tag{5.41}$$

where we have used the identity $\lim_{s\to 0^+} \frac{1}{x-is} = \mathcal{P}\frac{1}{x} + i\pi\delta(x)$. The above equation can be written in a more familiar form if we use Fermi's Golden rule and introduce the transition probabilities

$$W^0_{kk'} = \frac{2\pi}{\hbar}|\langle k|H'|k'\rangle|^2 \delta(E_k - E_{k'}) = 2\pi N \frac{|\phi_{kk'}|^2}{V^2} \delta(\omega_{kk'}). \tag{5.42}$$

With this the transport equation takes the usual form

$$\boxed{-ieE^0_\alpha \frac{\partial \rho_k}{\partial k_\alpha} = \frac{V}{(2\pi)^3} \int d^3k' (W^0_{k'k} f_k - W^0_{kk'} f_{k'}).} \tag{5.43}$$

This is what we set out to obtain (compare with the Boltzmann equation in chapter 2).

5.5 Why and how are the traditional kinetic theories justified?

From the Kohn-Luttinger formalism we observe that the off-diagonal elements of the Fourier-transformed electron-impurity interaction Hamiltonian are suppressed as compared to its diagonal elements when ensemble average over the impurities is performed. This in turn affects diagonal and off-diagonal elements of the impurity averaged density matrix. In the leading order we observe that the traditional transport equation is the end result. Thus, the formalism provides a kind of justification, within the restrictions of the model, of the fundamental assumption of the "random phase." For further details (including the higher order corrections) the reader is referred to [1].

5.6 Summary points

1. The traditional Boltzmann equation is based on many assumptions and approximations (refer to Chapter 2). The random phase assumption is one of them.

2. When we compute statistical ensemble average of an observable, we need the diagonal as well as the off-diagonal elements of the density matrix $\langle \mathscr{O} \rangle = tr(\rho\mathscr{O})$. But the traditional Boltzmann equation deals with the diagonal elements of the density matrix. The neglect of the off-diagonal matrix elements is based on the RPA.

3. The Kohn-Luttinger program, in a simple model of electron-impurity scattering, shows that the off-diagonal elements are actually suppressed when the ensemble average is performed over a random distribution of impurities. Thus, the formalism provides a basis for the traditional Boltzmann equation.

Bibliography

[1] W. Kohn and J. M. Luttinger, *Phys. Rev.* **108**, 590 (1957).

[2] L. Van Hove, *Physica* **21**, 517 (1955).

[3] R. C. Tolman, *Principles of Statistical Mechanics*, OUP, New York (1930).

[4] L. D. Landau and E. M. Lifshitz, *Statistical Physics*, Pergmon Press (1980).

Chapter 6

Strange metals: A survey

They [cuprates] highlight a major intellectual crisis in the quantum theory of solids, which, in the form of one-electron band theory, has been very successful in describing good metals (like Cu) but has proven inadequate for strongly correlated systems.

Damascelli, Hussain, and Shen*

6.1 General introduction to the problem

In the first chapter we learned that the Drude theory, although apparently able to account for the observed Wiedemann-Franz law, is not able to address the temperature dependence of electrical and thermal resistivities. The electrical resistivity of a simple metal, for example Au, Ag, Cu, etc., linearly increases with temperature at room temperature and exhibits T^5 law at low temperatures. The characteristic temperature that defines these two regimes is called the Debye temperature. Thus, the electrical resistivity of simple metals shows the following behavior:

$$\rho(T) \sim \begin{cases} T & : T >> \Theta_D \\ T^5 & : T << \Theta_D. \end{cases} \tag{6.1}$$

Rev. Mod. Phys. **75**, 473 (2003).

It became possible to understand the above temperature dependence of resistivity with the pioneering work of F. Bloch. As we learned in Chapter 2, Bloch was able to formulate the transport problem starting with the basic principles of the then newly discovered quantum mechanics. Bloch came up with a transport equation (now called the Bloch-Boltzmann equation) in which the collision term constitutes the scattering processes of electrons with lattice vibrations (the phonons). Thus, with Bloch's theory it became possible to understand the temperature dependence of electrical resistivity of metals in terms of the inelastic scattering of electrons with phonons. The theory and experiment are in reasonably good agreement, especially at temperatures above the Debye temperature.

However, at low temperatures the peculiarities of the Fermi surface matters and the simple-minded Bloch-Boltzmann theory based on the spherical Fermi surface has to be extended to take that into account. Philip Allen, William Butler, and collaborators have done a considerable amount of work regarding the application of the Bloch-Boltzmann theory to real metals taking the electronic band structure into account. Interested readers may refer to [1, 2], and references therein. The magneto-transport properties like magneto-resistance, Hall effect, etc., can be addressed within the Bloch-Boltzmann theory for the case of simple metals [3, 4].

The situation is not so straightforward in the case of strange metals (like normal states of high-T_c cuprates above the optimal doping). We will learn from the following discussion that the temperature variation of the electrical resistance cannot be understood in terms of Bloch-Boltzmann theory with electron-phonon scattering as a sole mechanism of momentum degradation. *Although there are many proposed theories that claim to address the T-linear resistivity, there is none that can address all the anomalous features observed over the whole phase diagram. The problem is to rule out the multiple explanations present in the literature and to come up with a single explanation which can address all the anomalous features in a coherent manner.*

Our plan for this chapter is the following. In the next section we undertake a thorough survey of the experimental situation on transport properties in the "normal" states of the cuprate superconductors. As our objective in the present book is the study of electronic transport, we are going to consider the anomalous behavior of (1) DC resistivity and (2) AC conductivity.† For the experimental situation related to these probes in other materials readers are referred to the literature [6]. [The list given is not complete and students are encouraged to search more literature using Internet.] At the end, the theoretical situation is reviewed.

6.2 A survey of the experimental situation

6.2.1 What are strange metals?

The strange metal regime occurs in many materials which shows (un)conventional superconductivity (superconductivity due to any non-phononic mechanism). The materials include cuprates, iron-based superconductors, heavy Fermion superconductors, organic superconductors, etc. Below we explicitly consider the case of cuprates (for other materials refer to [6]).

A cartoon phase diagram for cuprates is shown in Figure 6.1. The above phase diagram is a typical phase diagram as many features are universal among different families of cuprates. The phase diagrams represent various phases as a function of temperature and doping. The parent (undoped) cuprate materials (like La_2CuO_4, $YBa_2Cu_3O_7$, etc.) are Mott-insulators (more precisely, charge transfer insulators) with long range antiferromagnetc order.‡ With increasing doping, at a fixed temperature, these materials pass through a variety

†Strange metals show anomalous behavior in many other transport, thermal, and spectroscopic probes. Interested readers are referred to more comprehensive literature [11].

‡Cuprates are layered materials with perovskite structure. The active elements are CuO_2 planes which are separated from each other by charge reservoir layers. Introduction to

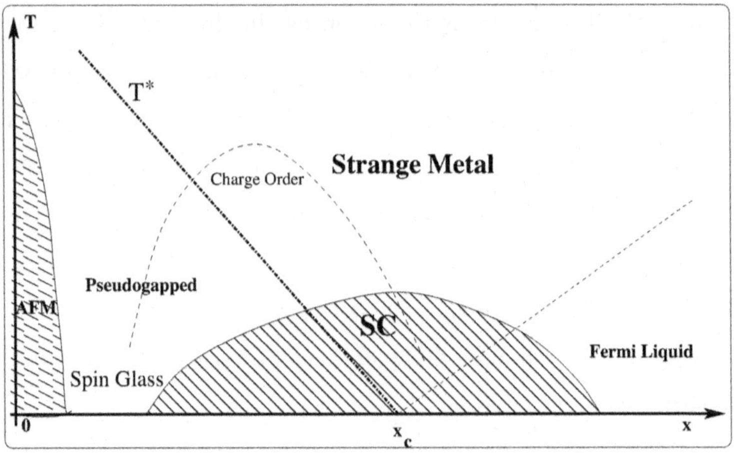

FIGURE 6.1: A typical phase diagram of cuprates.

of phases§ such as "melting" of the antiferromagnetic order, the pseudogapped state, the (un)conventional superconductivity, and Fermi liquid regime, etc. Superconductivity occurs in a dome-shaped region with maximum superconducting transition temperature at the optimal doping (Figure 6.1).

The strange metal regime, which is our topic of discussion, occurs above the superconducting dome. At higher doping (i.e., overdoping) strange metal regime gradually transforms to the Fermi liquid regime (there is no phase transition). And at lower doping (i.e., underdoping), it gradually transforms into a state that shows a partial gap (in some directions of the Brillouin zone) in the density of states called the pseudogapped regime (again, there is no phase transition, and it is called a cross-over).

As mentioned before, in the strange metal regime experimental probes

---

cuprate structure is very brief here. There is a large body of literature; interested readers can refer, for example, to [7, 8, 9].

§The doping here means changing the electron concentration in CuO_2 planes. If electrons are removed from CuO_2 planes, it is called the hole doping and if these are added, it is termed electron doping. Much like in the semiconductor physics.

show anomalous behavior. We start our survey with the simplest probe, i.e., the DC resistivity.

6.2.2 Anomalous behavior of DC resistivity

As cuprates are anisotropic materials their transport characteristics show a very wide range of variation with respect to the orientation of the crystal in an experimental setup. For example, the DC resistivity is 10^4 times higher along the c-axis (i.e., the inter-plane conductivity, ρ_c) than that in the ab-plane (i.e., CuO_2 planes, ρ_{ab}) in the most anisotropic single-layered cuprate $Bi_2Sr_2CuO_6$ $(Bi-2201)$. This anisotropy varies from material to material. In the most isotropic cuprate $YBa_2Cu_3O_7$ the ratio $\frac{\rho_c}{\rho_{ab}}$ is of the order of 10. The qualitative behavior of ρ_{ab} and ρ_c is also different. So we consider them separately. Let us start with the ab-plane DC transport.

6.2.2.1 ab-plane transport

6.2.2.2 Optimally doped cuprates

We have studied the DC resistivity of simple metals in Chapter 2 and we learned that it can be understood within the Bloch-Boltzmann kinetic theory. Cuprates show different behavior which does not seem to fall within the above-mentioned standard paradigm of the Bloch-Boltzmann theory. The single most important and well-documented observation in cuprates is the linear variation of ρ_{ab} with temperature at optimal doping (refer to Figure 6.2). It is evident from the figure that $\rho_{ab} \propto T$. In the figure only a sample of two cuprates (LSCO and Bi-2201) is shown, but the trend is typical among other cuprates. For $Bi-2201$ sample it shows T-linear behavior over a wide range of temperature from 7 K to 700 K. It is also observed that ρ_{ab} is an intrinsic property of the CuO_2 planes and weakly depends upon the character of the intervening charge reservoir layers. In Table 6.1 resistivities of various cuprates are tabulated [12]. Although there is a reasonable variation from

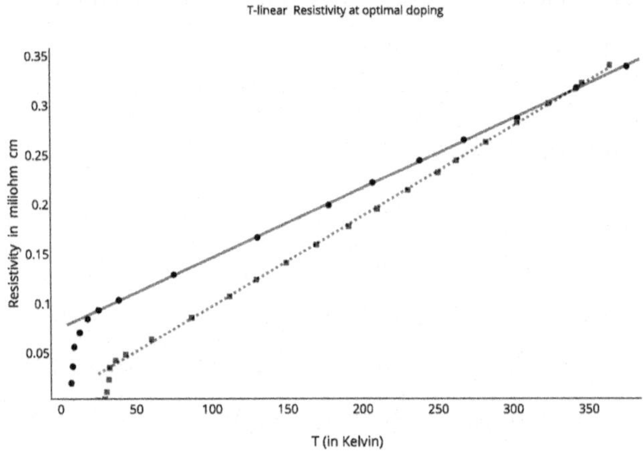

FIGURE 6.2: DC resistivity at optimal doping (a sample). Squares are the data points for $La_{2-x}Sr_xCuO_4$ at $x = 0.21$[10], and the dotted line is a linear fit to the data above the superconducting transition. Filled circles are the data points for $Bi-2201$ at optimal doping [11] and the solid line is a linear fit to the data.

TABLE 6.1: ab-plane resistivities of various cuprates per CuO_2-plane are of the same order of magnitude. This observation clearly shows that electronic transport is mainly controlled by the CuO_2-planes.

Compound	ρ_{ab} $(\mu\Omega cm)$	ρ_{ab} per layer
$YBa_2Cu_3O_{6.95}$	290	580
$La_{1.83}Sr_{0.17}CuO_4$	420	420
$Bi_2Sr_{1.61}La_{0.39}CuO_6$	500	500
$Bi_2Sr_2CaCu_2O_{8+\delta}$	280	560
$Tl_2Ba_2CuO_6$	450	450

material to material, ρ_{ab} per unit $Cu - O$ plane is roughly of the same order of magnitude. This has a clear message:

The scattering mechanisms, which are responsible for doping and temperature variation of ρ_{ab}, are intrinsic to the $Cu - O$ planes and have nothing much to do with the charge reservoir layers.

This reduces the theoretical problem considerably. It was first observed by Batlogg and this observation is contained in the "dogma" no. 1 in Phil Anderson's book on cuprate superconductivity [13]. The importance of these kinds of observations lies in their ability to reduce the complexity of the problems (as the problems related to cuprates are among the hardest problems in current condensed matter physics).

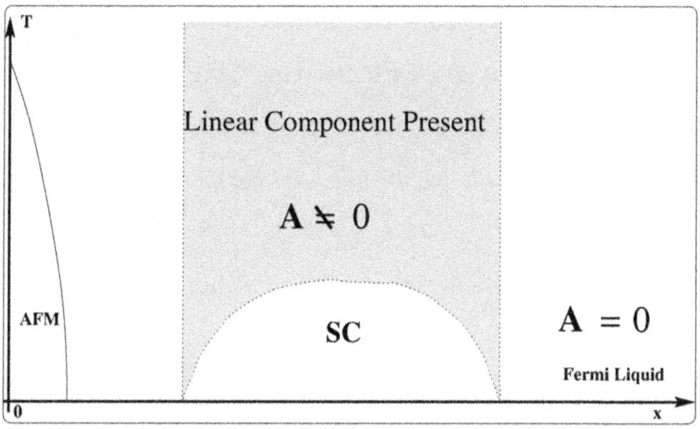

FIGURE 6.3: T-linear component of resistivity is always present in the regime above the SC dome. The shaded area implies the presence of the T−linear component. This points toward an important link between the mechanism of T-linear resistivity and the mechanism of Cooper paring in cuprates.

There is another important observation related to DC resistivity. As the materials are doped away from the optimal doping, either toward overdoping or toward underdoping, the DC resistivity ceases to have purely linear temperature dependence. If the resistivity is modeled as

$$\rho = AT + BT^2, \tag{6.2}$$

then $A \neq 0,\ B = 0$ at the optimal doping, but, away from optimal doping both the components (i.e., $A \neq 0,\ B \neq 0$) are present as far as one is above the superconducting dome in the phase diagram (refer to Figure 6.3). Taillefer and collaborators highlight this observation [14]. This key observation shows that a system goes into a superconducting state if and only if the DC resistivity has T-linear component in the "normal" state just above the superconducting dome. At overdoping when T_c goes to zero, the coefficient A also goes to zero. The magnitudes of A and the value of T_c are proportional to each other. Thus, there must be an important link between the mechanism that causes T-linear resistivity ($A \neq 0$) and the mechanism that causes superconductivity ($T_c \neq 0$). It is quite plausible that the T-linear resistivity originates from the scattering of electrons from some Fluctuating Magnetic Substratum (FMS)[¶] and the same FMS is responsible for Cooper paring when the temperature is brought below T_c. *At present, we do not have proper mathematical description of this possibility.*

The obvious question arises: what is the mechanism of the $T-$linear resistivity? One can argue that the T-linear resistivity is due to electron-phonon scattering (similar to that in metals). There are many points that stand against this argument, some of them are listed below:

1. At low temperatures $T << \Theta_D$ (where Θ_D is the Debye temperature), good metals show $\rho \propto T^5$. This behavior is never observed in the cuprates. Resistivity is perfectly T-linear, in $Bi-2201$ the linear regime extends from 7K to 700 K!

2. Resistivity of metals is shown in Figure 6.4(a). If one extrapolates the high temperature part $T > \Theta_D$ to low temperatures, then it is observed that the extended part cuts the temperature axis at $T \sim \frac{\Theta_D}{5}$.

[¶]As cuprates pass through many magnetic instabilities [14].

This is another important signature of electron-phonon scattering [2]. On the other hand, if one extrapolates the resistivity of cuprates (Figure 6.4(b)) it passes near the origin showing no appreciable intercept on the temperature axis (in high quality cuprate samples). This is an important observation that goes against the electron-phonon scattering as the mechanism of resistivity in cuprates.

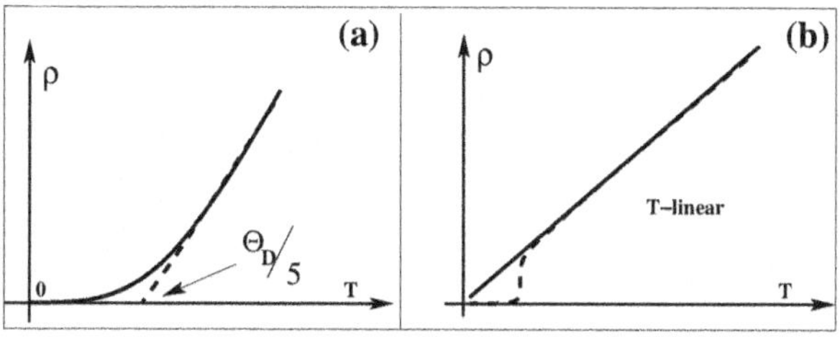

FIGURE 6.4: (a) Resistivity of metals and (b) resistivity of cuprates at optimal doping.

3. In the overdoped regime the resistivity behavior changes to $\rho \propto T^2$. This is a signature of the electron-electron scattering and one of the defining characters of a Fermi liquid. Neither T-linear nor T^5 behavior of resistivity is observed at overdoping. This has a clear message that electron-phonon scattering remains a sub-dominant mechanism even at overdoping.

4. The generalized Drude scattering rate $1/\tau(\omega)$ of cuprates shows no saturation behavior at $\omega \simeq \omega_D$ whereas in metals, where electron-phonon scattering is the dominant mechanism, this saturation is observed.

Thus, one can safely say that electron-phonon scattering is not the mechanism for resistivity in cuprates. Resistivity appears to be purely of electronic origin and the mechanism of Cooper pairing might be in the same vein.

There is another very peculiar feature with DC resistivity of cuprates. It does not show Mott-Ioffe-Regel (MIR) saturation limit (as it does in the case of simple metals). ab-plane DC transport at optimal doping is metallic in character. DC resistivity for 2-D $Cu - O$ planes can be written as[‖]

$$\sigma_{2D} = \frac{e^2}{2\pi\hbar}(k_F l), \tag{6.3}$$

where l is the electronic mean free path and k_F is the Fermi wave vector. The dimensionless parameter $k_F l$ determines whether the resistivity is metallic (if $k_F l >> 1$), or saturating the MIR bound (if $k_F l \sim 1$).[**] For rough estimation let us consider the case of $(La-Sr)CuO_4$ at optimal doping. At $T = 50K$, $\rho = 0.05\ m\Omega cm$. The resistivity per plane ρ_{2D} can be calculated from c-axis inter-plane distance $c \sim 6\mathring{A}$ (refer to problem (14) in Chapter 8). From Equation (6.3) the dimensionless parameter $k_F l$ turns out to be ~ 25. Thus, $k_F l >> 1$ in this case, and the transport is metallic in character. At higher temperature, say at $T = 800K$, it nearly saturates the bound as can be easily seen from the above formula. *Thus at optimal doping the transport remains metallic for most of the temperature scale.*

The situation is just opposite in the underdoped regime where σ_{2D} is substantially greater than MIR resistivity ρ_{MIR}, and transport becomes non-metallic ($k_F l < 1$). *Due to this reason the underdoped cuprates are termed "bad" metals.* We consider the temperature dependence of DC resistivity in the underdoped regime below followed by a discussion of overdoped cuprates and the c-axis transport.

6.2.2.3 Underdoped cuprates

In the underdoped cuprates there is an important energy scale associated with the opening of a partial gap in the density of states when the temperature is lowered below a characteristic temperature T^* $(T^* > T_c)$. This state with partial gap is called the pseudogapped state [15, 16].

It is observed that the resistivity is T-linear for $T > T^*$ and shows deviations from linearity below T^*. At T^* the slope of the resistivity curve shows

‖Refer to problem number (14) in Chapter 8.

**The condition $k_F \sim l$ implies that l is of the order of lattice constant $a \sim \frac{1}{k_F}$.

a break. Figure 6.5 shows the resistivity of $YBa_2Cu_4O_8$. One can observe a kink around $T^* \simeq 200K$ where the slope of the resistivity curve shows a change, and the resistivity is reduced below T^*. The temperature at which this is observed roughly coincides with the temperature at which a cross-over behavior is observed in spin-lattice relaxation rate in NMR experiments [17]. This reduction in resistivity is attributed to the reduction in scattering due

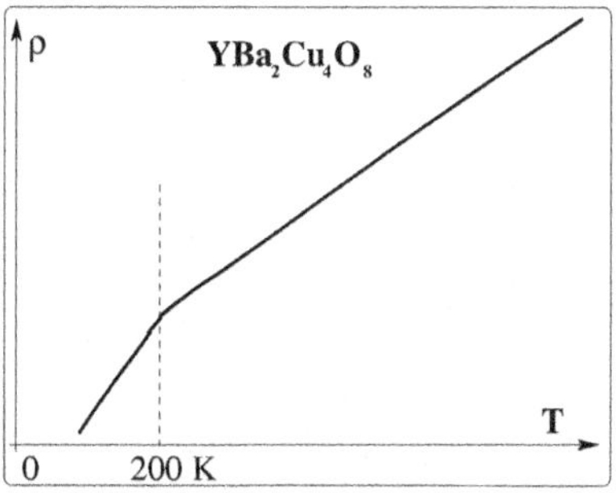

FIGURE 6.5: At $T \simeq T^*$ the resistivity shows a change in the slope which roughly coincides with what is observed in NMR experiments.

to the opening of the gap. In other words, it could be due to the removal of the spin scattering channel within the $Cu-O$ plane due to the opening of the pseudogap as the spin response becomes anomalous at $T = T^*$[18].

There is a nice scaling observed in the resistivity in the underdoped samples [19]. If one plots $\frac{\rho_{ab}(T)}{\rho_{ab}(T^*)}$ as a function of $\frac{T}{T^*}$ for different cuprates (Figure 6.6), the resistivity curves fall on the same line and show slope change at $T = T^*$. There is some deviation only below $\frac{T}{T^*} = 0.5$ (Figure 6.6). The scaling observed can have the following implication. If relaxation time approximation is used in the Boltzmann equation (Chapter 2) then resistivity turns out to be inversely proportional to the relaxation time τ, i.e., $\rho \propto \frac{1}{\tau}$, which further implies that

$$\frac{\tau(T^*)}{\tau(T)} = f\left(\frac{T}{T^*}\right). \tag{6.4}$$

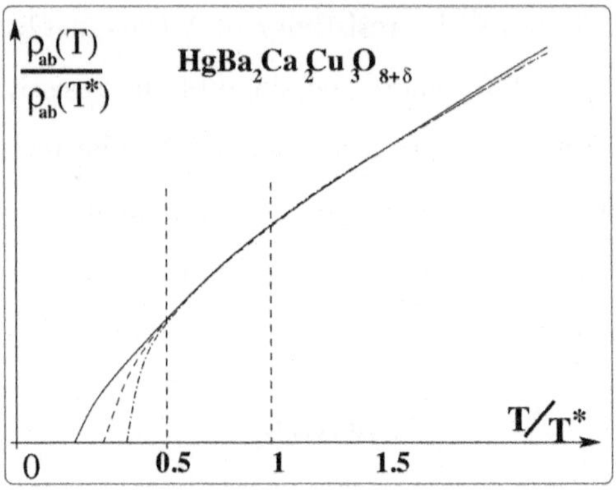

FIGURE 6.6: The scaling relation of resistivity in underdoped cuprates.

Here f is a universal function. *At present, we do not have a microscopic theory that can predict this kind of scaling.*

There is another kind of scaling observed at $T < T_c$. If superconductivity is suppressed by using very high pulsed magnetic fields, then it has been observed that

$$\rho \propto ln(1/T) \tag{6.5}$$

in the underdoped regime [20, 21]. The origin of this logarithmic divergence is not yet clear. Again, as mentioned before, underdoped cuprates are bad metals as they strongly violate the MIR bound (the measured resistivities are substantially greater than the MIR resistivity). The violation of this bound has a deeper implication. It implies that $l < a$. As we learned in Chapter 2, this implies that uncertainty in the momentum of the quasiparticles is greater than $\frac{1}{a} \sim k_F$, i.e., the zone momentum. Thus, momentum does not remain a good quantum number and the very basis of the quasiparticle picture is lost. In other words, the use of the Bloch-Boltzmann kinetic theory for the underdoped cuprates is not justified and one has to rely on other approaches like memory function formalism or Kubo's approach (cf. Appendix 8.3).

6.2.2.4 Overdoped cuprates

ab-plane DC transport in the overdoped cuprates can be understood within the Landau Fermi liquid theory. A line is usually drawn from the end of the superconducting dome at overdoping toward the upper right corner of the phase diagram and the regime below that putative line (Figure 6.1) shows Fermi liquid behavior where the DC resistivity shows T^2 behavior. Also as we will learn from the c-axis transport that in this regime the 2D character of the $Cu-O$ layers is lost and a 3D character emerges due to substantial c-axis charge coupling.

There are two main viewpoints regarding the transition from strange metal regime to Fermi liquid metal regime. In one view [8, 13] there is no real transition from strange metal regime to Fermi liquid regime and it is termed a gradual cross-over with the same electronic ground state. In the other viewpoint [22] it is believed that the nature of the ground state itself changes in a fundamental way and there is a cross-over. Both viewpoints are debated [22].

6.2.2.5 c-axis transport

Thus far we have discussed the ab-plane DC transport and we learned that there are many universal trends at all dopings. Next we will review the c-axis DC transport which is marked by lesser universal trends and has sample to sample variation in some doping range. As discussed at the beginning of this section, the cuprates are very anisotropic materials with c-axis resistivities $10-to-10^4$ times the ab-plane resistivities. We also learned that at optimal doping in the ab-plane the resistivity nearly saturates the MIR bound and violates it in the underdoping. The c-axis DC transport violates the MIR bound strongly both at underdoping and at optimal doping due to the above-mentioned reason. *Thus the c-axis transport is non-metallic.* Here non-metallic means $l < a$ and as we argued before, momentum does not remain a good quantum number. Thus, theories based on the Bloch-Boltzmann equation (at underdoping and optimal doping) have no justification and one has to adopt incoherent tunneling theories [23].

The picture is somewhat comfortable at overdoping where resistivity is substantially less than MIR limit and the ratio $\frac{\rho_c(T)}{\rho_{ab}(T)}$ becomes temperature independent implying the onset of coherent c-axis transport with a 3D character of the electronic ground state. Thus, resistivities can be addressed within the premises of the Bloch-Boltzmann theory.

The underdoped side transport is non-metallic and ρ_c shows an insulating character below $T < T^*$. It coincides with the slope changing feature in the ρ_{ab} at $T \sim T^*$. A plausible reason for higher ρ_c below T^* could be the formation of the spin-singlets that block the c-axis transport (refer to Anderson [13]). With this overview of the DC transport we next consider the AC transport, i.e., IR/optical conductivity, along similar lines.

6.2.3 Anomalous behavior of AC conductivity

6.2.3.1 ab-plane transport

As discussed in Chapter 1, AC conductivity is "extracted" from reflectivity/absorptivity measurements using ellipsometry. Optical "constants" are obtained using the standard Kramers-Kronig (KK) analysis and Hagen-Rubines type low frequency extrapolation of the data [24, 25]. We learned that simple metals show Drude behavior with $\sigma(\omega) \sim \frac{1}{\omega^2}$ at high frequencies and Drude peak at $\omega = 0$ whose width scales as $\frac{1}{\tau}$. We will learn that cuprates do not obey the Drude behavior in some parts of their phase diagram. Next we consider various doping regimes separately.

6.2.3.2 Optimally doped cuprates

There is a remarkable universality observed in the optimally doped cuprates for the optical conductivity $\sigma(\omega)$. For good samples of YBCO [26] optical conductivity, just above the superconducting dome, is best described by

$$\sigma(\omega) \sim \frac{1}{\omega}, \qquad for\ \ k_B T \lesssim \omega \lesssim 1\ eV. \tag{6.6}$$

This is the defining character of the strange metal regime (Figure 6.7(a)). This behavior is observed in a narrow region just above the optimal dop-

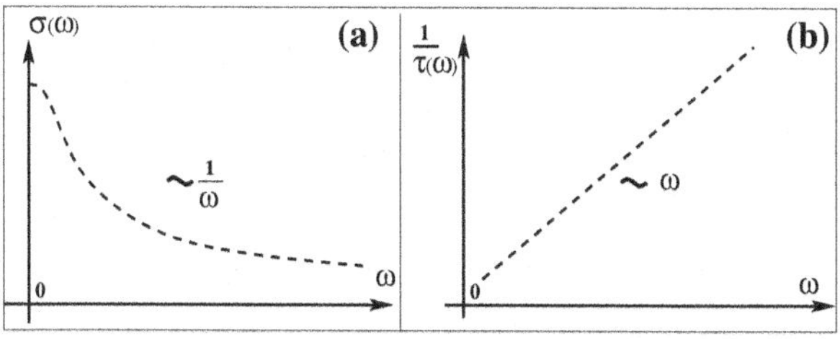

FIGURE 6.7: The defining characters of the strange metals regime.

ing where maximum T_c occurs. For the cuprate in the BISCCO family $Bi_2Sr_2Ca_{0.92}Y_{0.08}Cu_2O_{8+\delta}$ [27] this behavior, just above the optimal doping, takes the form

$$\sigma(\omega) \sim \frac{1}{\omega^{0.65}}, \qquad for\ \ k_B T \lesssim \omega \lesssim 1\ eV. \tag{6.7}$$

In $Bi_2Sr_2CaCu_2O_8$, $\sigma(\omega) \sim \frac{1}{\omega}$. A very low temperature superconductor $Bi_2Sr_2CuO_6$ ($T_c = 5K$) also shows the similar behavior ($\sigma(\omega) \sim \frac{1}{\omega}$). There are other numerous examples [9]. One can write this non-Drude form in a general way

$$\sigma(\omega) \sim \frac{1}{\omega^{\alpha}} \tag{6.8}$$

where α is a fraction of the order of one. But for the majority of the cuprates it is roughly one. In the next section we will study some theories that attempt to address this. Optical conductivity data is generally analyzed using Memory function (MF) formalism or the Generalized Drude (GD) formalism [9]:

$$\sigma(z) = i\frac{\chi_0}{z + M(z)}. \tag{6.9}$$

As we studied in Chapter 4 that this formula (also called the Mori formula) can be derived rigorously from the Kubo formula for dynamical conductivity, and can be applied to any system with arbitrarily strong electron-electron or electron-phonon interactions.[††] Thus, optical conductivity data for strongly correlated systems can be analyzed using MF or GD formalism. The MF has two parts: (1) the real part of it is related to the mass enhancement factor and (2) the imaginary part is related to the scattering rate called the generalized Drude scattering rate. [‡‡]

By fitting the experimentally observed conductivity with conductivity from the GD formula, the scattering rate and the mass enhancement factor can be extracted. Figure 6.7(b) shows the frequency dependence of the scattering rate (also called the generalized Drude scattering rate) obtained form the experimental conductivity above the optimal doping. The linear dependence of $\frac{1}{\tau(\omega)}$ on frequency is quite striking and it is termed another defining character of the strange metal regime. The real part of the MF, i.e., the mass enhancement factor, also shows frequency dependence albeit weaker than $\frac{1}{\tau(\omega)}$, and it reduces with increasing frequency. The high energy tails of $\sigma(\omega)$ are immune to temperature (the high frequency tails do not show much dependence on temperature (Figure 6.8). Thus, the experimental data leads to the following scaling laws:

$$\sigma(\omega) \propto \frac{1}{\omega}, \quad and \quad \frac{1}{\tau(\omega)} \propto \omega, \tag{6.10}$$

which are quite universal and are termed the AC analogue of the T-linear DC resistivity. Thus, the whole body of the literature on the ab-plane optical conductivity can be rationalized with

$$\boxed{\frac{1}{\tau(\omega)} \sim max\{k_B T, \ \omega\}.} \tag{6.11}$$

This pivotal observation is one of the most elementary requirements of any theoretical framework which claims to address all the anomalous features observed in the strange metal regime of cuprates.

[††] Only the light-matter interaction is treated as perturbation (i.e., the linear response).

[‡‡] In literature the MF is also called the optical self energy.

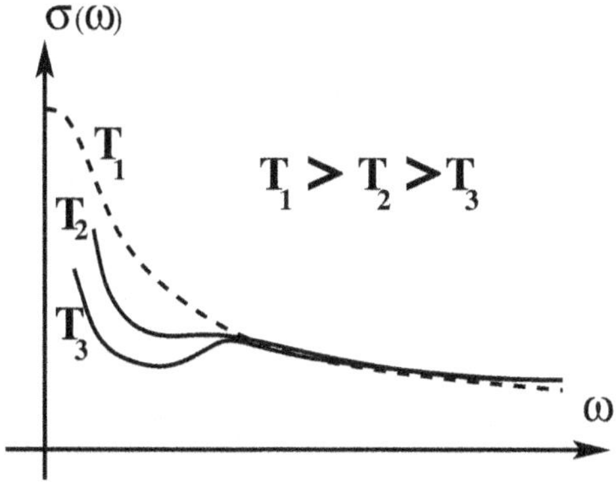

FIGURE 6.8: The high frequency part of the conductivity is temperature independent.

6.2.3.3 Underdoped cuprates

The situation is much more complex in the underdoped regime, where in non-superconducting state, two new energy scales come into play: (1) pseudogap energy scale, and (2) the energy scale associated with the mysterious Mid-Infrared band (MIR). These energy scales affect the optical spectrum in a very complex way. We learned that the DC resistivity shows a pronounced change of slope when the temperature is brought below T^* (Figure 6.5). An equivalent signature is seen in optical conductivity for $T < T^*$. The low frequency conductivity shows a dip and the spectral weight shifts from the low frequency Drude peak to mid-infrared frequencies. Figure 6.9(a) depicts the schematic behavior of the optical conductivity. As the temperature is reduced below T^* a dip-like feature appears at around $\omega \sim 500\ cm^{-1}$. This feature is a universal one and is observed in many families of cuprates [9]. As the doping level is increased at a fixed temperature, this dip-like feature weakens and disappears at overdoping. The doping trend of this Dip feature follows the T^* line drawn in the phase diagram (Figure 6.1). There is a consensus in the literature that this dip-like feature around $\sim 500\ cm^{-1}$ in the underdoped cuprates is due to the opening of the pseudogap.

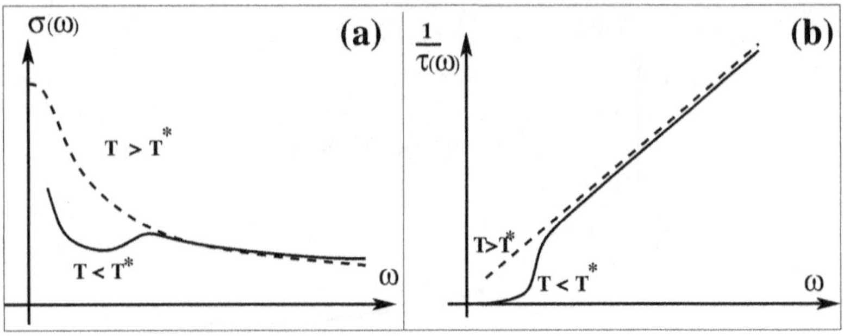

FIGURE 6.9: (a) At $T < T^*$ a gap-like feature appears in the optical conductivity with a dip at $\omega \sim 500\ cm^{-1}$. (b) A corresponding signature appears in the generalized Drude scattering rate when the temperature is brought below T^*.

The other energy scale is associated with the Mid-InfraRed band (MIR). It is well known that the undoped cuprates (parent materials) are Mott insulators or more precisely charge transfer insulators. The charge transfer gap of $\sim 1\ eV$ is seen in the optical conductivity at zero doping, where the conductivity is negligible below $\sim 1\ eV$ and shows a pronounced shoulder above $1\ eV$ [Figure (6.10), the dashed line]. The high frequency part ($> 2\ eV$) is roughly doping independent and thus it is not shown. The schematic diagram shown in Figure 6.10 is for $(La - Sr)_x CuO_4$. But a similar trend is also observed in other cuprates. When the doping is increased, spectral weight shifts from the high frequency part ($\omega > 1.0eV$) to the low frequency part ($\omega < 1.0eV$), i.e., into the charge transfer gap. The most important feature of the structure in the charge transfer gap region is a peak at around $\omega \sim 0.5\ eV$ [Figure (6.10), bump in the solid line]. This peak structure is quite robust and is seen in many cuprate families. *This peak structure is famously known as the MIR band.* As the doping is further increased, the MIR band weakens in magnitude and progressively shifts to lower frequencies and finally at overdoping it merges with the increasing spectral weight of the Drude peak. This happens at doping $x = 0.2$ where it is not possible to observe the MIR peak.

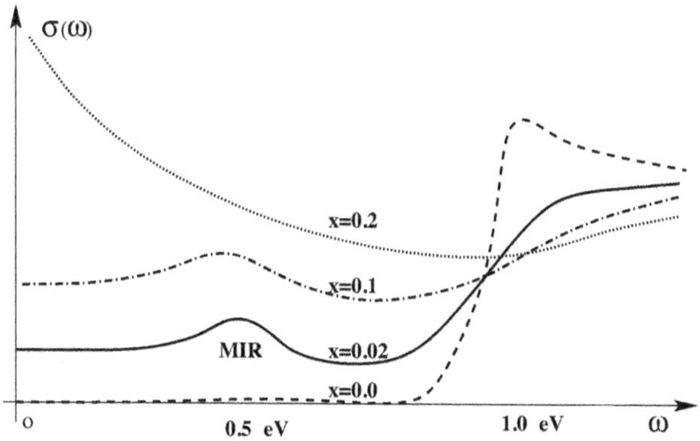

FIGURE 6.10: The famous MIR band appears as the material is doped away from the insulator state. With increasing doping this feature shifts to lower frequencies and disappears at overdoping.

As the other features of the "normal" state of cuprates remains poorly understood, the MIR band also remains a mystery. There are many views on its origin and we summarize them in Figure 6.11.

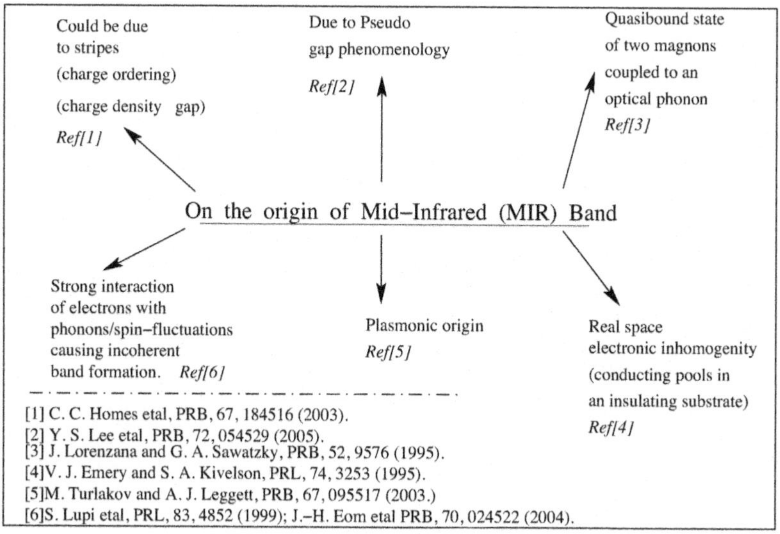

FIGURE 6.11: The mystery of the MIR band and its plausible explainations.

6.2.3.4 Overdoped cuprates

In overdoped cuprates, below that putative line (thin dashed line in Figure 6.1), the nature of charge transport is similar to that of a Fermi liquid. The optical conductivity shows Drude peak and Drude law tails ($\sigma(\omega) \sim \frac{1}{\omega^2}$). However, as mentioned before, the role of electron-phonon scattering remains sub-dominant.

6.2.3.5 c-axis transport

The pseudogap phenomenon is most clearly observed in the c-axis optical conductivity. Figure 6.12 depicts a schematic c-axis IR conductivity of $YBa_2Cu_3O_{6.7}$. For $T > T^*$ the c-axis optical conductivity is temperature and frequency independent (almost flat spectrum). As the temperature is lowered below T^* a gap-like feature (depression) begins to appear. Depth of this gap-like feature increases with decreasing temperature (Figure 6.12). However, its width (along the frequency axis) is quite robust and temperature independent. The observed gap is roughly of the order of 300 cm^{-1}. As the temperature is lowered below T^*, the lost spectral weight at low frequencies shifts to higher frequencies ($\omega \gtrsim 1eV$). The temperature at which the depres-

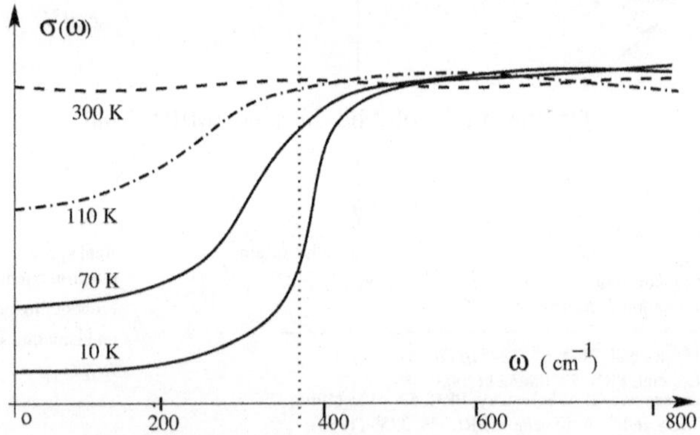

FIGURE 6.12: The c-axis IR conductivity showing the opening of a pseudogap below $T^* \simeq 300K$.

sion appears (i.e., the pseudogap temperature T^*) roughly coincides with that inferred from the slope changing feature in NMR spin susceptibility measurements [9]. It also matches with the temperature at which a slope changing feature is observed in the ab-plane DC resistivity. This shows the consistency of the observations. The c-axis conductivity above $500cm^{-1}$ is roughly temperature independent. Another important feature that has been observed is that the magnitude (width) and the depth of the depression seen does not change when the system goes into the superconducting state when the temperature is lowered below T_c. This implies that the superconducting transition does not affect this aspect of the pseudogap phenomenon.

At the optimal doping in the strange metal regime c-axis conductivity is flat, i.e., it is temperature and doping independent. In experiments, many optical phonon modes are observed which are understood. The important point that we have to recognize is that the electronic contribution to the c-axis conductivity in the strange metals regime is both temperature and doping independent.

In the overdoped regime (beyond the strange metals regime) it is observed that the c-axis conductivity shows the emergence of a metal-like response in which the IR conductivity is no longer temperature and frequency independent. The conductivity shows the Drude peak and decreases with increasing frequency. This feature is more prominent at lower temperatures [28].

6.3 A survey of the theoretical situation

In the previous section we have surveyed the main experimental anomalies in the electronic transport probes observed in the strange metal regime. As mentioned before, the strange metal observed in the normal states of cuprate superconductors shows anomalous behavior in many other transport, thermal, and spectroscopic probes. Interested readers can consult more comprehensive reviews [11]. In the present section we are going to review the two prominent theoretical frameworks that attempt to address these anomalies (our focus will remain on AC and DC conductivity). In fact, there are many viewpoints regarding the strange metals with novel ideas, and the literature is vast (on the theory side also). We are going to mention them and will try to organize them. Then two viewpoints will be discussed in some detail. Let us first mention the prominent ideas:

1. The Marginal Fermi Liquid (MFL) theory (Varma and collaborators [29]).

2. Nearly antiferromagnetic Fermi liquid based theories (David Pines and collaborators [30]).

3. The "hot spots" and "cold spots" heuristics (Millis, Rice, and collaborators [31]).

4. Quantum criticality based theories (Sachdev and collaborators [32]).

5. Anderson's Hidden Fermi Liquid (HFL) theory [33].

6. Ideas from soft condensed matter physics and particle physics (Balatsky and collaborators [34]).

We are going to discuss the MFL theory and the HFL theory in some detail as these theories are comparatively more successful in addressing a multitude of experimental probes. But before we do that it is useful to organize the literature into two ways.

1. Fermi-liquid versus non-Fermi-liquid based theories

2. Microscopic versus Phenomenological theories

The nearly antiferromagnetic Fermi liquid theory of Pines and collaborators, the "hot" and "cold" spots heuristics, etc., are Fermi liquid based theories, in which the ground state of the strange metal is treated as a Fermi liquid with well-defined quasiparticles. Refer to Appendix 8.3 in the last chapter.

The other set of theories, such as MFLT of Varma and collaborators, HFLT of Phil Anderson, theories based on quantum critical point hidden under the superconducting dome and strange metal as its finite temperature projection, etc., are non-Fermi liquid theories. In these theories the quasiparticle concept in the Landau sense is not valid. For example, in MFLT the quasiparticle weight Z goes to zero logarithmically at the Fermi surface and single-particle self energy proposed by it is anomalous (it differs from the Fermi liquid self energy, as discussed in detail below). In the HFLT of Phil Anderson, to account for the strong on-site Coulomb repulsion, the Gutzwiller projection method is used in the Hubbard model. The resulting Hilbert space (after Gutzwiller projection) is termed the projected Hilbert space where double occupancy on a given Cu site is prohibited. In the HFLT the elementary excitations of the unprojected Hilbert space (prior to projection) are not quasiparticles but pseudoparticles and they obey the rules of Fermi liquid theory (HFL ansatz). Quasiparticles pertain to the projected Hilbert space and these are the physical electrons (i.e., those which are excited by the applied electric field). We will discuss these ideas in detail below. In quantum criticality based theories, quantum critical fluctuations due to the quantum critical point generally hidden under the superconducting dome destroy the Landau quasiparticles.

The second classification based on the phenomenological versus microscopic theory is, to some extent, a subjective issue (in this complex field). The MFLT can be classified as a phenomenological theory, as the anomalous self energy which the theory proposes has its motivation in NMR and Raman scattering experiments (the original idea was to obtain the required input from a certain class of experiments and using that input, the idea was to propose the single-particle self energy and to obtain the output of the formalism. And then that output is compared with other experiments). Thus, MFLT can be classified as a phenomenological theory.

While the HFLT of Anderson can be classified as a microscopic theory as it starts from the "microscopic" Hubbard model and with the method of Gutzwiller projection it imposes the physical requirement of no double occupancy at any Cu site. Theory postulates the existence of pseudoparticles (excitations of the (un)projected Hilbert space) and quasiparticles (excitations of the projected Hilbert space). However, there is no mathematical formulation in which pseudoparticles and quasiparticles are natural outcomes (recall Dirac's beautiful mathematical theory of particles and anti-particles, i.e., relativistic quantum mechanics. There antiparticles are a natural prediction of the theory).

After describing these approaches, their problems are also discussed. As mentioned before, we have chosen MFLT and HFLT for detailed description as these theories are comparatively more successful in addressing the experimental probes. With this broad exposure, we start to review the marginal Fermi liquid theory.

6.3.1 The marginal Fermi liquid theory

The fundamental premise of the MFLT is based on the MFL hypothesis:

Over a wide range of momentum **q**, there exist excitations which make a contribution to both charge and magnetic excitation spectra $\xi(\mathbf{q},\omega,T)$ of the form [29]:

$$Im\xi(\mathbf{q},\omega,T) \sim \begin{cases} -N_0(\omega/T) & : \ \omega << T \\ -N_0(sign\omega) & : \ T << \omega << \omega_c \end{cases} \quad (6.12)$$

Evidently, it has a scale invariant form such that $Im\xi \propto f(\omega/T)$. In the above equation N_0 is the density of states per unit volume and ω_c is a high frequency cut-off. The single-particle retarded self-energy from the above spectrum of charge and magnetic excitations takes the form:

$$\begin{aligned} \Sigma(\mathbf{k},\omega) &= \Sigma_1(\mathbf{k},\omega) + i\Sigma_2(\mathbf{k},\omega) \\ &= \lambda(\omega \log(\frac{x}{\omega_c}) - i\frac{\pi}{2}x), \end{aligned} \quad (6.13)$$

where $x = max(|\omega|, T)$,[§§] and λ is a coupling constant. The above equation is the defining equation of the MFLT.

The MFLT differs from the conventional Fermi liquid theory in a very fundamental way. In Fermi liquid theory, the single-particle self energy at zero temperature has the form: $Re\Sigma \sim \omega$ and $Im\Sigma \sim -\omega^2$. The quasiparticle picture ($|Im\Sigma| << |Re\Sigma|$ at $k_B T << \epsilon_F$) is valid (Appendix 8.3). The single-particle Green's function can be written as

$$G(\mathbf{k}, \omega) = \frac{1}{\omega - \epsilon_{\mathbf{k}} - \Sigma(\mathbf{k}, \omega)} = \frac{z_{\mathbf{k}}}{\omega - E_{\mathbf{k}} + i\Gamma_{\mathbf{k}}} + G_{incoh}, \tag{6.14}$$

where $z_{\mathbf{k}} = (1 - \frac{\partial Re\Sigma}{\partial \omega})^{-1}_{\omega = E_{\mathbf{k}}}$ is the quasiparticle weight and it is bounded by $0 < z_{\mathbf{k}} < 1$. In contrast to it, in MFLT, $z_{\mathbf{k}}$ shows a logarithmic dependence on quasiparticle energy (i.e., $z_{\mathbf{k}} \sim 1/\ln\left|\frac{\omega_c}{E_{\mathbf{k}}}\right|$). Near the Fermi surface (i.e., when $E_{\mathbf{k}} \to 0$) it goes to zero, logarithmically. Thus, in a MFL there are no well-defined quasiparticles (in the Landau sense). Consequently, the Green's function of a MFL is entirely incoherent. Here it differs fundamentally from a conventional Fermi liquid. In a Fermi liquid the spectral function shows a sharp peak around the quasiparticle energy (with width much smaller than the quasiparticle energy). In a MFL the quasiparticles are much wider with a substantial weight in the wings of the spectral function (see Figure 1 in the first reference in [29]). These are the fundamental principles of the MFLT.

With the proposed MFL self-energy, many anomalous properties of the strange metal can be addressed (but not all). Let us first consider the resistivity. We have seen in the experimental section that the resistivity is dead linear at the optimal doping (with a small constant term that could be due to impurities, and not important to us here). In [29] the authors used the Drude formula

$$\rho = (\omega_p^2 \tau(T))^{-1}, \tag{6.15}$$

and by fitting with experiment they observed that

$$\frac{1}{\tau(T)} \sim T. \tag{6.16}$$

---

§§For example, x can be of the form $x = \sqrt{\omega^2 + \pi^2 T^2}$.

This is in accord with their proposed self-energy which in the DC limit is given by

$$Im\Sigma(\mathbf{k},\omega) \simeq \frac{1}{\tau} \simeq -\lambda\frac{\pi}{2}T \propto T. \tag{6.17}$$

Thus, their model self-energy is consistent with the experimental result of T-linear resistivity. But there are two subtle points that the authors of [29] do not discuss:

1. As we learned in Chapters 1 and 2 the Equation (6.15) for resistivity ($1/\rho = \sigma = \frac{ne^2\tau}{m}$) can be obtained from the Drude model or from the Boltzmann equation under relaxation time approximation. These transport theories are based on the concept of quasiparticles. If the quasiparticle concept ($z_{\mathbf{k}} \to 0$) breaks down in a MFL (like in the strange metals regime), then these theories have no justification for their applicability in a marginal Fermi liquid, and within these premises, it is not possible to justify the Drude formula, i.e., Equation (6.15). However, the Kubo theory can be applied even if the quasiparticles are not well defined (as in the present case) and the DC conductivity can be written as

$$\sigma_{\mu\nu} = \beta\int_0^\infty dt\langle J_\nu(0)J_\mu(t)\rangle. \tag{6.18}$$

   To obtain the Drude formula from the above Kubo formula one has to make a fundamental assumption that current-current correlation function must be an exponentially decaying function with a temperature dependent decay rate:

$$\langle J(0)J(t)\rangle \sim e^{-t/\tau(T)}. \tag{6.19}$$

   To the best of the author's knowledge, this form (also called the Markovian form) has never been shown in a marginal Fermi liquid setting.

2. The second subtle point concerns the scattering rate and the imaginary part of the single-particle self-energy (Equation (6.17)). In a general setting the transport scattering rate $\frac{1}{\tau_{tr}}$ differs from the quasiparticle

decay rate $\frac{1}{\tau} \sim Im\Sigma$. The former is weighted by angle depended scattering (small angle scattering is not sufficient in the degradation of momentum/current, and the large angle scattering is much more effective, $\frac{1}{\tau} = \int d\Omega M(\Omega)(1 - \cos\theta)$, refer to Chapter 4). The authors discuss this point in the second reference in [29]. The authors argue that Σ_2 in MFLT is momentum independent and thus Σ_2 can be substituted for $\frac{1}{\tau_{tr}}$.

Thus, given the above points, T-linear resistivity is in accord with MFL self-energy.

Next comes the AC conductivity $\sigma(\omega)$. We have seen that the AC conductivity can be rationalized using generalized Drude formalism with frequency and temperature dependent scattering rate:

$$\frac{1}{\tau(\omega)} \sim \begin{cases} \omega & : \hbar\omega > k_B T \\ k_B T & : \hbar\omega < k_B T \end{cases} \tag{6.20}$$

This is directly in accord with the proposed MFL self energy where the imaginary part of it can be written in the above form. Although the above trend is general, in the case of $Bi_2Sr_2Ca_{0.92}Y_{0.08}Cu_2O_{8+\delta}$, the conductivity shows $1/\omega^{0.65}$ dependence. It could be a peculiarity of the system. However, at present the exact reason is not known [35]. Similarly other probes like tunneling conductance, nuclear relaxation rate, specific heat and thermal conductivity, etc., can be addressed (see for details [29]).

Within MFLT one can understand the violation of the Mott-Ioffe-Regel (MIR) bound at high temperatures in the optimally doped regime, and in the underdoped regime. Since MFLT implies the absence of quasiparticles, the criterion on which MIR bound is founded does not apply.

Also, the authors in their very first paper on it [29] argue that their proposed charge and magnetic excitation spectrum can address the superconducting pairing as pairing kernel $v(\mathbf{q}) \times Re(1/\epsilon(\mathbf{q}, \omega))$, where $v(\mathbf{q})$ is the bare electron-electron interaction, shows considerable independence on $\mathbf{q}$ (according to their hypothesis which proposes no or weak momentum dependence of charge and magnetic excitations). Thus, they proposed s-wave pairing to be a likely candidate. However, later experiments showed d-wave pairing.

It is true that MFLT is directly grounded in the experimental facts and has some success. However, there are some subtle problems. For example, Hall resistivity and Hall angle measurements cannot be addressed within the present premises. We will learn while discussing the hidden Fermi liquid theory of Anderson that to address Hall angle measurements one has to invoke Fermi liquid scattering for Hall scattering rate. There should be a component of Fermi liquid self-energy that, somehow, is not important for electrical transport but becomes important under the application of magnetic field. Thus, to address Hall angle measurements further physics is needed. Also the resistivity and conductivity is only addressed at the optimal doping. At overdoping, it is well known that the physics can be understood within the Fermi liquid theory; resistivity is quadratic with temperature ($\rho \propto T^2$). It would be nice to have a complete picture where MFL self-energy changes to FL self-energy with doping. At underdoping, the physics is further complicated by pseudogap effects. Thus, it is important first to focus on optimal and overdoping. Clearly much work is needed to address these problems.

With this brief introduction to MFLT we next move over to the HFLT.

6.3.2 The hidden Fermi liquid theory

The central idea on which HFLT is based pertains to the strong on-site repulsion between the doped holes. The physics of the CuO_2 planes is mainly controlled by the Hubbard U, and the effective low energy model of CuO_2 planes is the $t-J$ model in which U is renormalized to infinity. Using Gros-Rice transformation, the Hubbard Hamiltonian can be transformed into

$$H_{t-J} = \mathcal{P}(\sum_{i,j,\sigma} t_{ij} c^{\dagger}_{i\sigma} c_{j\sigma})\mathcal{P} + \sum_{i,j} J_{ij} S_i.S_j, \tag{6.21}$$

which is famously known as the t-J model [36, 37]. The projector $\mathcal{P} = \prod_i (1 - n_{i,\uparrow} n_{i,\downarrow})$ in the above equation is called the Gutzwiller projector. The physical meaning of this operator, as acting on the kinetic energy term in the above Hamiltonian, is that it "projects out" all the doubly occupied sites in all possible configurations of the doped holes on Cu sites. Central to the hidden Fermi liquid theory are the two kinds of Hilbert spaces: one is associated with the projected states and the other is associated with the unprojected states.

HFLT consists of the following assertions. The true excitations of the system are the excitations in the unprojected Hilbert space. Let us denote them by $c_{i,\sigma}$, $c^{\dagger}_{i,\sigma}$. These are called the pseudoparticles (PPs). In contrast, the quasiparticles (i.e., electrons) are the excitations of the projected Hilbert space and obey the Gutzwiller projection. These are the excitations when an external electric field is applied. Let us denote them by "hat" operators $\hat{c}_{i,\sigma}$, $\hat{c}^{\dagger}_{i,\sigma}$. Gutzwiller projection implies:

$$\mathcal{P} \underbrace{c^{\dagger}_{i\sigma}}_{pseudoparticles} = \underbrace{\hat{c}^{\dagger}_{i,\sigma}}_{physical\ \ electrons} . \tag{6.22}$$

The basic ansatz of the HFLT is the following: The PPs (the excitations of the unprojected Hilbert space) obey the rules of the Fermi liquid theory, i.e., their spectral function exhibits sharp peaks and has finite quasiparticle weight z, while, the excitations of the projected Hilbert space (QPs or physical electrons) constitute a non-Fermi liquid, in that they have broad spectral function peaks and vanishing quasiparticle weight z. And there is no adiabatic continuity between PPS and QPs.

When an external stimulus is applied to the system (like an electric field) it is the physical electrons that couple with the field and get acceleration. However, to dissipate momentum to the lattice, they must themselves decay to the excitations of the unprojected Hilbert space, i.e., to PPs (as electrons are not the true excitations of the unprojected Hilbert space). *Thus the process of transferring momentum from the field to the lattice becomes a two-step process. As we will observe, it is the first process which causes the strange metal behavior.*

The relevant Green's function that one needs to describe the momentum relaxation is the physical electron Green's function (i.e., with hat operators):

$$\begin{aligned} G^{holes}_{proj,i\sigma} &= \langle 0|\hat{c}^{\dagger}_{i,\sigma}(t)\hat{c}_{i,\sigma}(0)|0\rangle \simeq \langle 0|c^{\dagger}_{i,\sigma}(t)c_{i,\sigma}(0)|0\rangle \\ &\times \langle 0|c_{i,-\sigma}(t)c^{\dagger}_{i,-\sigma}(t)c_{i,-\sigma}(0)c_{i,-\sigma}(0)c^{\dagger}_{i,-\sigma}(0)|0\rangle \end{aligned} \quad (6.23)$$

The last factor $G^* = \langle 0|c_{i,-\sigma}(t)c^{\dagger}_{i,-\sigma}(t)c_{i,-\sigma}(0)c_{i,-\sigma}(0)c^{\dagger}_{i,-\sigma}(0)|0\rangle$ has the crucial physics [33]. The first factor in the above equation is for PPs and describes the conventional Fermi liquid (the basic ansatz of the HFLT). It is the second factor G^* which is all important. Anderson argues that the G^* pertains to a classic problem in metals, i.e., the problem of X-ray edge singularities [33, 38]. The basic idea of the connection is that in the classic X-ray edge singularity problem a high energy photon kicks out a deep electron in the core of an atom in the sample. The hole generated due to the removal of the electron then acts as a scatterer for other conduction electrons. But the potential of the hole generated is not static in time. As the hole is refilled by an electron, its potential abruptly goes to zero. Thus one is dealing with a scattering problem but with potential that is NOT constant in time. This problem, scattering by fluctuating potential, was solved in the 1970s [38]. *It turns out that the Green's function for that deep holes shows power law behavior.* This is highlighted as the crux of the whole matter.

Anderson argues that the dynamics of the extra holes (when the system is hole doped) at $Cu-$sites shows a behavior similar to the fluctuating potential behavior as seen in the X-ray edge problem. The dynamics of the extra holes create a transient potential much like the potential of the deep hole in X-ray edge problem. Thus the Green's function of the doped holes has power law behavior which explains the strange metal behavior as described below.

A finite temperature extension of the problem by Yuval and Anderson [39] shows that the decay rate of the deep hole is proportional to temperature. By invoking the above connection, HFLT asserts that the rate at which the physical electrons (induced by an applied electric field) decay to pseudoparticles is also proportional to temperature. With this ansatz the authors of [40] explain many observed properties of the strange metal regime. We consider the important case of T-linear resistivity.

As mentioned before, when the electric field is applied, it is the physical electrons that are excited. These electrons must decay to PPs for the momentum to be transferred to the lattice. It turns out that at optimal doping the first process acts like a bottleneck (Figure 6.13). Let us define that the time scale on which QPs decay to PPs is τ_1 and the time scale on which momentum is transferred to the lattice by PPs is τ_2 (i.e., by umklapp processes). The timescale τ_1 is the longest time scale at optimal doping (the bottleneck).

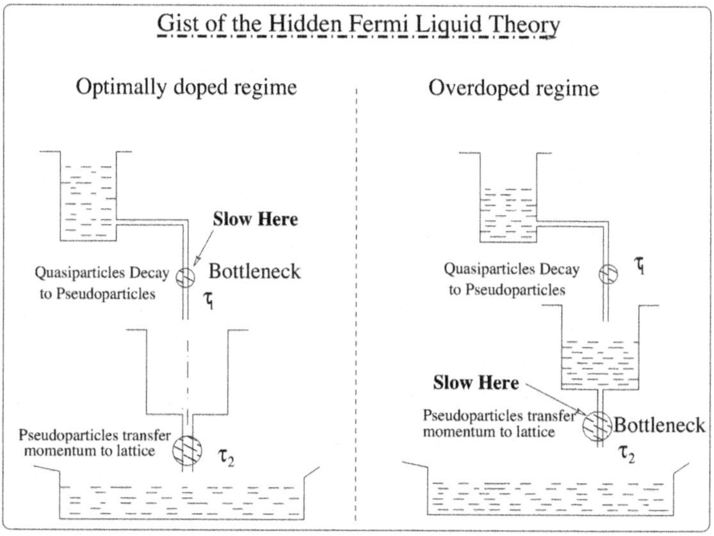

FIGURE 6.13: The "bottleneck" mechanism of the HFLT.

Thus transport lifetime is $\tau_{tr} = \tau_1 + \tau_2 \simeq \tau_1$ and the resistivity is given by $\rho = \frac{m}{ne^2}\frac{1}{\tau_{tr}} \simeq \frac{m}{ne^2}\frac{1}{\tau_1} \propto T$ (as discussed before in the context of X-ray edge problem).

In the overdoped regime, the opposite of the above is true. The bottleneck happens at the second step (Figure 6.13). Due to increased doping, thus increased effective bandwidth W_{HFL} for PPs, their scattering rate becomes smaller ($\Gamma \sim \frac{1}{W_{HFL}}$[40]). Thus timescale τ_2 becomes the largest timescale, and it controls the net momentum transfer to the lattice, or, $\tau_{tr} \simeq \tau_2 \propto \frac{1}{T^2}$. The time scale τ_2 is inversely proportional to the square of the temperature as PPs obey the rule of the Fermi liquid theory, and the momentum is transferred to the lattice by umklapp processes.

Thus within HFLT it is possible to address the temperature dependence of resistivity, and its evolution with doping, i.e., the change of T-linear resistivity at optimal doping to T-square resistivity at overdoping. It is a smooth crossover in HFLT (rather than the abrupt change of behavior).

The other anomalous feature of the strange metal regime is the Hall angle ($\cot\Theta_H = \frac{1}{\omega_c \tau_H}$) which can also be addressed within the HFLT. Casey and Anderson argue that the applied magnetic field does not excite the quasiparticles (i.e., electrons) of the projected space (as magnetic field commutes with the projection operator), rather, it excites the PPs of the unprojected space. This stands in sharp contrast to what an electric field does to the system. Thus it is the time scale τ_2 which is all important in Hall angle measurement ($\tau_H = \tau_2$) and $\cot\Theta_H = \frac{1}{\tau_H} \propto T^2$. Therefore, the Hall angle measurement can be rationalized within HFLT. The authors of [40] go further. From fitting of the Hall angle measurement with their theory they obtain the effective bandwidth W_{HFL}. When this effective bandwidth is used in the DC resistivity, a very good agreement between the theory and the experiment is observed. However, this "self-consistency" is shown only at overdoping (not at optimal doping).

On similar lines, the power law behavior of the AC conductivity can be addressed [40]. The relevant Green's function that goes into the Kubo formula is the "X-ray edge motivated" Green's function G^*, which shows the power law behavior [41]. The other anomalous features of the strange metal regime can be addressed on similar lines [41].

The picture painted by the HFLT is by no means perfect. Theory consists of many ansatz and assumptions. Although it has a-posteriori justification (through consistent description of Hall angle and resistivity measurements), it would be nice to start out with a microscopic Hamiltonian, and with systematic mathematical deductions, to reach to all the premises of the theory. By this route, PPs and QPs will be the natural outcomes. This state has not arrived yet and much work is needed. Also, the authors have shown self-consistency only at overdoping. A detailed analysis of all the experimental probes within HFLT over a wide range of doping and temperatures is lacking.

Another important question should be asked: What happens when the system becomes superconducting? As the superconductive state¶¶ is quite "conventional" in cuprates, in that it shows Bogolyubov quasiparticle peak. It would be nice to have a mathematical theory that can explain the transformation of "pseudoparticle-quasiparticle" picture of the strange metal regime into the Bogolyubov quasiparticle picture of the superconducting state.

In conclusion, the HFLT, although it has many appealing assertions and some successes, needs much further work for other probes, and especially it needs a microscopic theory to justify its assertions.

There are other theoretical scenarios that we are going to discuss in brief in the next subsection.

6.3.3 A brief discussion of other approaches

As we learned, the MFLT has problems in addressing the Hall resistivity and Hall angle measurements, and to address the doping evolution of DC resistivity and AC conductivity, further work is needed. It would be nice to have a microscopic justification of the MFL phenomenology. In the HFLT, the ansatz that the excitations of the unprojected Hilbert space are pseudoparticles which obey the rules of the Fermi liquid theory, although receives a-posteriori justification through the reconciliation of DC resistivity data with Hall angle data, does not have a mathematical proof. It would be nice to have a mathematical formalism where one starts with a microscopic Hamiltonian

¶¶Not the mechanism of superconductivity!

and systematically reach, with controlled mathematical deductions, to the premises of the HFLT. Also, self-consistency has only been shown at overdoping. It is desirable to investigate it at optimal doping. A detailed temperature and doping evolution of other experimental probes would establish the HFLT in a more concrete position where drawbacks will be more clear, and the scope of further improvement will enhance. Quantum criticality/ADS-CFT based and "hot", "cold" spots based theories face problems while addressing, for example, the IR/optical conductivity data (i.e., the power law behavior of optical conductivity and most importantly its doping evolution).

In Quantum Criticality (QC) motivated approaches, the strange metal is a finite temperature projection of zero temperature Quantum Critical Point (QCP). QCP is situated on a zero temperature nonthermal parameter line (like doping or magnetic field), and the finite temperature wedge shaped region constitutes the strange metal regime (Figure 6.1). For concreteness, consider that we are dealing with a spin system and let ξ be the ground state spin-coherence length scale [42]. Let us consider strange metal behavior in view of quantum criticality motivated scenarios. For length scales $<< \xi$, wavefunction of spins has entangled form, and for length scales $>> \xi$ it has product form. The shape of the QC regime is motivated in the following way. Let v_s be the spin-wave velocity. Then the thermal de-Broglie wavelength is given by $l_{th} \sim \frac{\hbar}{mv_s} \sim \frac{\hbar v_s}{k_B T}$. The wedge shaped QC regime is given by the condition $\xi > l_{th}$ or $T > \frac{\hbar v_s}{k_B \xi}$. Since ξ is a function of nonthermal parameter x (like doping), and it decreases with increasing x (away from x_c), it implies that $T(x)$ ($> \frac{\hbar v_s}{k_B \xi(x)}$) is an increasing function of x (away from x_c). Thus it is not hard to observe that it covers fan-shaped or wedge shaped regions known as the QC regime (Figure 6.1). Next comes the pivotal assertion. QC motivated approaches assign a special relaxation time to quantum critical regime, and the strange metal regime is identified with the QC regime. This special relaxation time scale is called the "quantum critical thermal equilibration time" or the

"Planckian dissipation limit":

$$\tau_{pdl} = c_{eq} \frac{\hbar}{k_B T}. \tag{6.24}$$

It is postulated to be the universal timescale in the QC regime, and it is the shortest possible relaxation time given by fundamental bounds like uncertainty principle [42]. If strange metal is a QC regime, then the Plankian dissipation can explain very high resistivity and its T-linear behavior ($\rho \propto \frac{1}{\tau_{pdl}} \propto T$). However, there are problems and the field has not reached a mature point. It is difficult to address the generalized Drude scattering rate $\frac{1}{\tau(\omega)}$ (consequence of power law AC conductivity) and the Hall angle measurements with $\frac{1}{\tau_H}$ in a unified, microscopic quantum criticality motivated setting. Also there are other anomalous features like NMR relaxation rates, Raman scattering with constant background, etc., that need to be addressed in a QC setting. Clearly more work is needed. Also in quantum criticality based approaches, a QCP is thought to be hidden under the SC dome. If that is the case, then there should be a sign of a diverging correlation length scale in some physical observables. However, such a correlation length scale is not seen in the experiments.

In conclusion, there are many issues that need to be addressed before we can have a complete picture. The field is definitely not closed yet and much work needs to be done. We would like to end this chapter with the following quote:***

The deep issue is that for a highly correlated fluid, the interactions are large and so likely cannot be treated using any fundamentally perturbative approach which starts with a free particle description. There is a well developed and extremely successful theoretical solution of this problem applicable to 1D and quasi-1D electron fluids based on "Bosonization," but no such approach exists in higher dimensions...

***B. Keimer, S. A. Kivelson, M. R. Norman, S. Uchida, and J. Zaanen, *Nature* **518**, 179 (2015).

Bibliography

[1] P. B. Allen and W. H. Butler, *Physics Today*, December issue, 44-49 (1978).

[2] P. B. Allen et al., *PRB*, **34**, 4331 (1986).

[3] I. M. Lifshitz and M. I. Kaganov, *Sovt. Phys. Uspekhi,* **87**, 805 (1966).

[4] G. D. Mahan, *Many-Particle Physics*, 3rd ed. Springer (2000).

[5] T. Yoshida et al., *J. Phys. Soc. Jpn.* **81**, 011006 (2012); M. Fujita et al., *J. Phys. Soc. Jpn.* **81**, 011007 (2012); T. Timusk and B. Statt, *Rep. Prog. Phys.* **62**, 61 (1999); J. R. Kirtley, *Rep. Prog. Phys.* **73**, 126501 (2010); E. Dagotto, *Rev. Mod. Phys.* **66**, 763 (1994); O. Fischer et al., *Rev. Mod. Phys.* **79**, 353 (2007).

[6] M. R. Norman, *Science* **332**, 196 (2011)[and references therein]; D. N. Basov and A. V. Chubukov, *Nat. Phys.* **7**, 272 (2011)[and references therein]; Yoshichika Onuki et al., *J. Phys. Soc. Jpn.* **73**, 769 (2004).

[7] P. A. Lee, N. Nagaosa, X. G. Wen, *Rev. Mod. Phys.* **78**, 17(2006).

[8] P. W. Anderson, *Int. J. Mod. Phys. B* **25**, 1 (2011).

[9] D. N. Basov and T. Timusk, *Rev. Mod. Phys.* **77**, 721 (2005).

[10] N. E. Hussey et al., *Phil. Trans. R. Soc. A* **369**, 1626 (2011).

[11] S. Martin et al., *Phys. Rev. B* **41**, 846 (1990).

[12] N. E. Hussey, Normal state transport properties, in *Handbook of High-Temperature Superconductivity: Theory and Experiment*, Ed. by J. R. Schrieffer, Springer (2007).

[13] Phil Anderson, *The Theory of Superconductivity in the High-T_c Cuprates*, Princeton Series in Physics (1997).

[14] L. Taillefer, *Ann. Rev. Cond. Matt. Phys.* **1**, 51 (2010).

[15] J. R. Cooper and J. W. Loram, *J. Phys.* **6**, 1 (1996).

[16] T. Timusk and B. Statt, *Rep. Prog. Phys.* **62**, 61(1999).

[17] B. Bucher et al., *Phys. Rev. B* **70**, 2012 (1993).

[18] T. Ito et al., *PRL* **70**, 3995 (1993).

[19] A. Carrigton et al., *Physica* **234c**, 1 (1994).

[20] G. S. Boebinger et al., *PRL* **77**, 5417 (1996).

[21] S. Ono et al., *PRL* **85**, 638 (2000).

[22] Feature article containing views of several pioneers, "Towards a complete theory of high-T_c", *Nat. Phys.* **2**, 139 (2006).

[23] N. Kumar and A. M. Jayannavar, *Phys. Rev. B.* **45**, 5001 (1992).

[24] J. S. Toll, *Phys. Rev.* **104**, 1760 (1956).

[25] F. Wooten, *Optical Properties of Solids*, Academic Press (1972).

[26] Z. Schlesinger et al., *PRB* **41**, 11237 (1990).

[27] van Der Marel et al., *Nature* **425**, 271 (2003).

[28] J. Schützmann et al., *PRL* **73**, 174 (1994).

[29] C. M. Varma et al., *PRL* **63**, 1996 (1989); E. Abrahams and C. M. Varma, *PNAS* **97**, 5714 (2000).

[30] D. Pines, *Physica C* **235**, 113 (1994); P. Monthoux et al., *PRB* **46**, 14803 (1992); P. Monthoux et al., *PRB* **49**, 4261 (1994); T. Moriya et al., *J. Phys. Soc. Jpn.* **59**, 2905 (1990).

[31] L. B. Ioffe et al., *PRB* **58**, 11631 (1998); R. Hlubina et al., *PRB* **51**, 9253 (1995).

[32] M. A. Metlitski, *PRB* **91**, 115111 (2015); S. Sachdev and B. Keimer, *Phys. Today*, February issue, 29 (2011).

[33] P. W. Anderson, *Nat. Phys.* **2**, 626 (2006); P. W. Anderson, *PRB* **78**, 174505 (2008).

[34] Jian- Huang She et al., arXive:1509.00991 (2015).

[35] M. R. Norman and A. V. Chubukov, *PRB* **73**, 140501(R) (2006).

[36] C. Gros, R. Joynt, and T. M. Rice. *PRB* **36**, 381 (1986).

[37] P. W. Anderson et al., *J. Phys.:Cond. Matt.* **16**, R755 (2004).

[38] P. Nozieres and C. T. de Dominicis, *Phys. Rev.* **178**, 1097 (1969); S. Doniach and M. Sunjic, *J. Phys. C.* **3**, 285 (1970).

[39] P. W. Anderson and G. Yuval, *PRL* **23**, 89 (1969).

[40] P. A. Casey and P. W. Anderson, *PRL* **106**, 097002 (2011); P. A. Casey and P. W. Anderson, *PRB* **80**, 094508 (2009).

[41] P. A. Casey, Ph.D. Thesis, Princeton Physics (2010).

[42] S. Sachdev and B. Keimer, *Phys. Today,* February issue, 29 (2011). B. Keimer et al., *Nature* **518**, 179 (2015).

Chapter 7

Electronic transport theories from simple to strange metals: A summary

The purpose of this brief chapter is to summarize the various electronic transport theories, starting from the very first theory of electrical conduction in metals by Paul Drude in 1900, and ending with recent theories designed to address the anomalous electronic transport in the strange metals regime of (un)conventional superconductors.

In the last 100 years or so, starting with the the pioneering work of Drude in 1900, great achievements have been made in elucidating the nature of electronic transport in a variety of materials. The first biggest achievement (after Drude theory) came with the advent of quantum mechanics when Felix Bloch (in the early 1930s) treated electrons in metals quantum mechanically and rewrote the Boltzmann equation by considering the scattering of electrons off the lattice vibrations (the Bloch-Boltzmann program). It was made possible by the advent of quantum statistics: Fermi-Dirac for electrons and Bose-Einstein for phonons. Felix Bloch was able, in one fell swoop, to rectify the problems of the Drude theory (i.e., in accounting for the temperature dependences of electrical and thermal conductivities) and at the same time established the foundations of the modern electronic theory of metals. The apparent success of the Drude theory in accounting for the Wiedemann-Franz law was also understood, i.e., his erroneous estimations for electronic velocity and electronic heat capacity, and their cancellation in the expression for heat conductivity (refer to Chapter 1). It was understood that quantum mechanical electronic velocity (i.e., the Fermi velocity) was two orders of magnitude more, and using quantum mechanics the electronic heat capacity was less by a similar order (refer to Chapter 1).

After Bloch's pioneering contribution, the Bloch-Boltzmann theory became the benchmark theory for the electronic transport in standard metals, and still for simple cases it enjoys the same status. At the heart of the theory is the fundamental concept of quasiparticles. The extension of the Boltzmann equation to electrons in metals is not straightforward (refer to Chapter 2). It is well known that Ludwig Boltzmann discovered the equation bearing his name while trying to understand the origin of the second law of thermodyamics from molecular collisions, and it describes the time evolution of the molecular distribution function due to applied external fields and internal inter-molecular interactions (Chapter 2). Thus, it was used to understand classical dynamical systems. But electrons in metals have to be treated quantum mechanically (to address specific heat and other transport anomalies, etc.). Thus, the very basis for the applicability of the Boltzmann equation (which was designed for classical particles) to a quantum system (electrons in a metal) is questionable.

As one can argue that the Bloch-Boltzmann equation has had great success in accounting for the temperature dependence of electrical and thermal resistivities, thus it receives a-posteriori justification (much like equal-a-priori probability hypothesis of statistical mechanics). But there is a very powerful theoretical argument for the justification of the applicability of the Boltzmann equation to electrons in metals. The justification rests on the quasiparticle concept of the Landau Fermi liquid theory (in short FLT) pioneered by Lev Landau (refer to Appendix 8.3). Landau showed that the low energy excitations of an interacting electronic system can have well-defined quantum numbers (like momentum remains a good quantum number in metals), and the system of strongly interacting electrons (via Coulomb interaction) can be mapped to a system of weakly interacting quasiparticles (one-to-one correspondence). Thus, with the quasiparticle concept, i.e., with well-defined momentum, the distribution function $f(\mathbf{r}, \mathbf{p}, t)$ can be defined on phase space whose cells are much bigger than $\hbar^3$, so that the uncertainty principle is not violated (refer to Chapter 2). There is another very subtle assumption which we did not mention in the second chapter. This is called the assumption of "random phase." It amounts to neglecting the off-diagonal elements of the density matrix. The diagonal elements correspond to the occupation probabilities of various states, i.e., the distribution function in the Boltzmann equation. There is an elegant and mathematically rigorous formalism due to Kohn and Luttinger which shows that the neglect of the off-diagonal elements of the density matrix in the traditional kinetic equations can be justified. It is the ensemble average over a random distribution of impurities that preferentially suppresses the off-diagonal elements of the density matrix, thus providing a justification of the random phase assumption (Chapter 5).

Up to the middle of the last century ($\sim$ 1950s), the Bloch-Boltzmann was the only theory to deal with nonequilibrium and transport problems. It was applied to a variety of situations including the anomalous skin effect [1].

In an elegant mathematical work, Reuter and Sondheimer in 1948 successfully applied the Boltzmann equation to understand anomalous skin effect in metals [2]. Rudolf Peierls extended the Bloch-Boltzmann equation to the regime where nonequilibrium nature of the phonon distribution becomes important (i.e., cases where phonon drag becomes important)[3].

The decade from 1950 to 1960 was very fruitful for nonequilibrium statistical mechanics and electronic transport theories. Many pioneers contributed, mainly, Ryogo Kubo and his colleagues from Japan; D. N. Zubarev, N. N. Bogolyubov, Leonid Keldysh and their colleagues from Russia; Rudolf Peierls and colleagues from the UK; R. Zwanzig, Ted Holstein, W. Kohn, J. Luttinger, L. Kadanoff and colleagues from the US; Radu Balescu, Ilya Prigogine and their colleagues from Belgium, etc. There are many useful ideas, however, the linear response theory as formulated by Kubo became very popular, and still it remains one of the leading approach. With the advent of Linear Response Theory (LRT) it became possible to express the transport coefficients in terms of the appropriate correlation functions. Thus mathematically it is quite different from the Bloch-Boltzmann equation. The semiclassical limit expressions obtained from LRT coincide with those obtained from the Bloch-Boltzmann equation approach. Most importantly, the LRT is fully quantum in nature as no semiclassical approximation is needed in principle, and it can also be applied where Landau Fermi liquid theory breaks down, i.e., where the quasiparticle concept is not justified. However, due to the mathematical difficulties progress in this direction is limited. We have observed this situation for the strange metals (in the previous chapter). But there is an other important difference between the Boltzmann equation approach and the LRT. The linear response theory, as its name suggests, is only applicable to weak nonequilibrium situations (like Ohmic regime, etc.), whereas the Boltzmann equation can be applied even to a situation which is arbitrarily far removed from equilibrium (like "hot" electrons in island metal films in micro-electronics applications [4]). However, the extensions of the LRT do exist in the literature (refer, for example, to the works of Stratonovich [5]), but again, with increasing mathematical complexity. The LRT, in the 1980s and 1990s, also saw its applications in mesoscopic physics where size of system becomes an important length scale. In the present treatment we have not discussed the electronic transport in mesoscopic systems. There are many good references in the literature and interested readers may refer to [6, 7].

In the same decade (1950 to 1960) another important formulation, to deal with nonequilibrium situations, was put forward by Mori. As we have shown in Chapter 4, from the Kubo formula for transport coefficients, Mori's expression in terms of the memory function can be obtained rigorously. Thus, both the formulations are identical. However, the Mori formulation has an added advantage. The transport coefficients can be written in the form of the Drude expression where constant Drude scattering rate is replaced by a temperature and frequency dependent imaginary part of the memory function (the generalized Drude scattering rate). We showed in Chapter 4 that the memory function in the impurity scattering case in metals reduces to a constant at very low frequencies ($\omega << \epsilon_F$) (i.e., to Drude scattering rate). Thus one can say that the Mori formulation is an extension of the Drude theory to situations where Drude scattering rate cannot be treated as constant and starts to depend on temperature and frequency. Due to this reason it is also known as generalized Drude formalism.

In the 1960s, the Bloch-Boltzmann theory for the electron-phonon scattering was revisited by Ted Holstein (refer to his elegant review article [8]). With his rigorous mathematical formulation he verified Landau's quasiparticle criterion of the applicability of the Boltzmann equation to electrons in metals (namely, $\frac{1}{\tau} << \epsilon_F$). His starting point was the Kubo LRT. Using sound speed to Fermi velocity ($\frac{v_s}{v_F}$) as a perturbation parameter, and using the then-newly discovered quantum field theoretic techniques (diagrammatic perturbation theory), he was able to obtain all the results of the traditional transport theory. The transport equation that he obtained is now called the Holstein-Boltzmann equation. Thus he was able to show that from the Kubo formula, one can obtain the Bloch-Boltzmann equation and in doing so all the physical requirements of the kinetic theory become clear. Due to its technical complexity this topic is not discussed in this introductory text.

In the same decade, a parallel approach was put forward by Kadanoff, Martin, Baym, and collaborators. These approaches again make use of the field theoretic diagrammatic techniques and the machinery of Green's functions. The end results are the transport equations similar to that of the Bloch-Boltzmann theory. In the 1980s, these approaches were revisited by Hänsch and Mahan. We have not gone into these approaches in the present text; however, interested readers can consult [9].

We return back to the memory function formalism which is the central topic of this book. In the 1970s, the MF approach of Mori was applied for the computation of the dynamical conductivity of metals by Wölfle and Götze (WG)[10]. We have reviewed their work in detail in Chapter 4. Through a perturbative expansion they were able to express the memory function in terms of current-current correlation function that was computed perturbatively assuming weak electron-impurity or weak electron-phonon scattering. *An important point is that although the MF was computed perturbatively, the result is beyond the Born approximation.* As was shown in Chapter 4, the expression for the scattering rate does contain $(1 - \cos\theta)$ factor, which signifies the importance of large angle scattering in the transport scattering rate. We obtained this result from the Boltzmann equation using an iterative procedure in an isotropic medium that goes beyond the relaxation time approximation (i.e., in Chapter 2). The memory function approach is "automatically" beyond the relaxation time approximation. Also in the AC limit for electron-phonon scattering, the Holstein mechanism is correctly captured by the WG formalism, and in the DC limit the Bloch-Boltzmann results for DC resistivity are obtained.

It is nice to obtain the known results of the Bloch-Boltzmann theory using MF formalism. But much more important is the possibility to apply the MF formalism to situations where Landau quasiparticle concept breaks down and one has to deal with a non-Fermi liquid state. This kind of situation occurs in the strange metal regime of (un)conventional superconductors.

In Chapter 6 we reviewed the observed anomalous electronic transport properties of strange metal seen in the "normal" states of the cuprate high-T_c superconductors. We learned that similar strange metal behavior is also observed in other unconventional superconductors (like heavy Fermion and iron-based superconductors, etc.). Among many anomalous features we observed that the DC resistivity at optimal doping is T-linear over a very wide temperature regime (over two decades of temperature). It was argued that it may not be due to phonons as T^5-law at low temperatures has not been observed and there is no intercept on the temperature axis of the extrapolated high temperature resistivity. We observed that the strange metals violate the Mott saturation limit. The AC conductivity shows non-Drude or power law tails at high frequencies [i.e., for simple metals $\sigma(\omega) \sim \frac{1}{\omega^2}$, and for strange metals $\sigma(\omega) \sim \frac{1}{\omega}$]. Similarly other probes show anomalous behavior [11]. There is a very peculiar link between the coefficient (A) of T-linear resistivity ($\rho = AT + BT^2$) and the magnitude of the transition temperature T_c. It turns out that $A \propto T_c$ and when there is no T-linear component in resistivity then there is no superconductivity [12]. *This implies a connection between the mechanism of T-linear resistivity and the mechanism of superconducting pairing.*

Returning back to the issue of anomalous electronic transport we observed that traditional theories (like Bloch-Boltzmann) are not adequate to address the strange metal regime. It is not like there are no theories for the strange metal regime, in fact, there are many novel ideas in the literature. In the previous chapter we classified them into two ways: (1) Fermi-liquid versus non-Fermi-liquid based theories, and (2) microscopic versus phenomenological theories. We learned that nearly antiferromagnetic Fermi liquid theory of Pines and collaborators, "hot" and "cold" spots heuristics, etc. are Fermi liquid based theories in which the ground state of the strange metal regime is treated as a Fermi liquid. The theories such as Marginal Fermi Liquid Theory (MFLT) of Varma and collaborators, Hidden Fermi Liquid Theory (HFLT) of Phil Anderson, theories based on quantum critical point hidden under the superconducting dome and strange metal as its finite temperature projection, etc. are (non)Fermi liquid theories. In these theories the quasiparticle concept in the Landau sense is not valid. For example, in MFLT the quasiparticle weight Z goes to zero logarithmically at the Fermi surface and in the HFLT the elementary excitations of the unprojected Hilbert space are not quasiparticles but pseudoparticles (we learned that in the HFLT the Gutzwiller projection is the governing step and the physical electrons pertain to the Gutzwiller projected Hilbert space where double occupancy on a given Cu site is prohibited). In quantum criticality based theories, quantum critical fluctuations destroy the Landau quasiparticles.

The classification based on the phenomenological versus microscopic theory, in this complex field, is a subjective issue to some degree. HFLT can be classified as the phenomenological theory as the anomalous self energy which the theory proposes has its motivation in Raman scattering experiments (the original idea was to obtain the required input for the proposed formalism from certain experiments, and to compare the output with other experiments). While the HFLT can be classified as a microscopic theory as it starts from the "microscopic" Hubbard model with the method of Gutzwiller projection to impose the physical requirement of no double occupancy at any Cu site, but again, there is no mathematical formulation in which pseudoparticles and quasiparticles are natural outcomes of a mathematical formalism (like Dirac's beautiful mathematical theory of particles and anti-particles, i.e., relativistic quantum mechanics).

In the previous chapter we reviewed these theories and observed that there are many issues that need to be addressed before we can have a complete picture. The field is definitely not closed yet and much work needs to be done. We observed that MFLT has problems in addressing the Hall resistivity and Hall angle measurements in the strange metal regime of cuprates, and further work is needed to address the doping evolution of DC resistivity and AC conductivity. In the HFLT, the ansatz that the excitations of the unprojected Hilbert space are pseudoparticles that obey the rules of the Fermi liquid theory, although receives a-posteriori justification through the reconciliation of DC resistivity data with Hall angle data, does not have a mathematical proof. It would be nice to have a mathematical formalism where one starts with a microscopic Hamiltonian and systematically reach, with controlled mathematical deductions, the premises of the HFLT. Also self-consistency has only been shown at overdoping. It is desirable to investigate it at optimal doping [13, 14].

Quantum criticality/ADS-CFT based and "hot", "cold" spots based theories face similar problems, for example, in addressing the IR/optical conductivity data (i.e., the power law behavior of optical conductivity and its doping evolution). Also in quantum criticality based approaches, a QCP is thought to be hidden under the SC dome. If that is the case, then there should be a sign of a diverging correlation length scale in some physical observables. However, such a correlation length scale is not yet seen in the experiments.

From our discussions in the previous chapter and in the present chapter we observe that much more work is needed to address the strange metal, both theoretically and experimentally. There is one point that the present author wishes to share with a reader who is a newcomer to the field. The reader has to deal with the complexity at two levels: (1) the complexity of the literature (the literature is extremely vast) and (2) the complexity posed by nature. This field of strange metals and unconventional superconductivity is by now about three decades old and thus it has a huge body of literature. The urgent need of time is to rule out what is not relevant and to focus on ideas that are directly coming from experiments and thus have some element of truth. Also to address these problems in a self-consistent way, a thorough knowledge of a variety of experimental probes is needed. It is not sufficient to address DC resistivity and Hall angle only, for example. A consistent picture must address all the probes like Raman scattering, AC/DC conductivity, Hall angle measurements, and other spectroscopic and thermal probes (with detailed temperature and doping evolution). This state has not arrived yet, and the "complete" picture is still lacking [15].

In the end, the author would like to add that the purpose of this text is best served if it is able to induce some excitement in newcomers. There are many difficult problems. The field is mature and remains open, and it should not happen that it is displaced by "fashion."

Bibliography

[1] A. B. Pippard, in *Optical Properties and Electronic Structure of Metals and Alloys*, edited F. Abeles (Noth-Holland, Amsterdam, 1966).

[2] G. E. H. Reuter and E. H. Sondheimer, *Proc. Roy. Soc. A*, **195**, 336 (1948); E. H. Sondheimer, *Adv. Phys.* **50**, 499 (2001)[reprinted].

[3] R. E. Peierls, *Quantum Theory of Solids*, Oxford University Press (1955).

[4] Navinder Singh, *Int. J. Mod. Phys. B* **24**, 1141 (2010).

[5] R. L. Stratonovich, *Nonlinear Nonequilibrium Thermodynamics I: Linear and Nonlinear Fluctuation-Dissipation Theorems*, Springer-Verlag (1992).

[6] S. Datta, *Electronic Transport in Mesoscopic Systems*, CUP (1995).

[7] Y. Imry, *Introduction to Mesoscopic Physics*, OUP (2001).

[8] T. Hosltein, *Ann. Phys. (N. Y.)* **29**, 410 (1964).

[9] W. Hänsch and G. D. Mahan, *Phys. Rev. B* **28**, 1902 (1983); R. E. Prange and L. P. Kadanoff, *Phys. Rev.* **134**, A566 (1964); L. P. Kadanoff and G. Baym, *Quantum Statistical Mechanics,* Benjamin, New York (1962).

[10] W. Götze and P. Wölfle, *Phys. Rev. B.* **6**, 1226 (1972).

[11] T. Yoshida et al., *J. Phys. Soc. Jpn.* **81**, 011006 (2012); M. Fujita et al., *J. Phys. Soc. Jpn* **81**, 011007 (2012); T. Timusk and B. Statt, *Rep. Prog. Phys.* **62**, 61 (1999); J. R. Kirtley, *Rep. Prog. Phys.* **73**, 126501 (2010); E. Dagotto, *Rev. Mod. Phys.* **66**, 763 (1994); O. Fischer et al., *Rev. Mod. Phys.* **79**, 353 (2007). [There are many more references that can

be searched with Internet search engines, and going into the references of the cited papers in the above list].

[12] L. Taillefer, *Ann. Rev. Cond. Matt. Phys.* **1**, 51 (2010).

[13] P. A. Casey and P. W. Anderson, *Phys. Rev. Lett.* **106**, 097002 (2011); P. A. Casey and P. W. Anderson, *Phys. Rev. B* **80**, 094508 (2009).

[14] P. A. Casey, Ph.D. Thesis, Princeton Physics (2010).

[15] B. Keimer et al., *Nature* **518**, 179 (2015).

Chapter 8

Supporting material and practice exercises

8.1 Appendix A: Proving $f^*_{\mu\nu}(\omega) = f_{\nu\mu}(\omega)$

First prove that $f_{\mu\nu}(-\omega) = f_{\nu\mu}(\omega)$. To see this consider $\phi_{\mu\nu}(t) = \int_0^\beta d\lambda \langle J_\nu(-i\hbar\lambda)J_\mu(t)\rangle$ or

$$\begin{aligned} \phi_{\mu\nu}(-t) &= \int_0^\beta d\lambda \langle J_\nu(-i\hbar\lambda)J_\mu(-t)\rangle = \int_0^\beta d\lambda tr(J_\mu(-t)e^{-\beta H}J_\nu(-i\hbar\lambda)) \\ &= \int_0^\beta d\lambda tr(e^{-iHt/\hbar}J_\mu(0)e^{-\beta H+\lambda H}J_\nu(0)e^{-\lambda H}), \end{aligned} \quad (8.1)$$

where the cyclic property of the trace operation is used, further, the integrand can be written as $tr(e^{-\beta H}e^{(\beta-\lambda)H}J_\mu(0)e^{-(\beta-\lambda)H}e^{iHt/\hbar}J_\nu(0)e^{-iHt/\hbar})$ or $\langle J_\mu(-i\hbar(\beta-\lambda))J_\nu(t)\rangle$. Thus,

$$\phi_{\mu\nu}(-t) = \int_0^\beta d\lambda \langle J_\mu(-i\hbar(\beta-\lambda))J_\nu(t)\rangle. \quad (8.2)$$

Shift the integration variable from λ to $\lambda' = \beta - \lambda$

$$\phi_{\mu\nu}(-t) = \int_0^\beta d\lambda' \langle J_\mu(-i\hbar\lambda)J_\nu(t)\rangle = \phi_{\nu\mu}(t). \quad (8.3)$$

Now, in the definition of the Fourier transform of the response function (Equation (3.53)) change ω to $-\omega$ and then shift the integration variable from t to

$-t$

$$f_{\mu\nu}(-\omega) = \int_{-\infty}^{+\infty} dt\phi_{\mu\nu}(-t)e^{i\omega t} = \int_{-\infty}^{+\infty} dt\phi_{\nu\mu}(t)e^{i\omega t} = f_{\nu\mu}(\omega). \quad QED. \tag{8.4}$$

Now, it is easy to prove that $f^*_{\mu\nu}(\omega) = f_{\nu\mu}(\omega)$. To see this, take the complex conjugate of the last equality in the above equation and proceed backward by shifting t to $-t$.

8.2 Appendix B: Computing $\langle(|H'_{kk'}|^2)^2\rangle_{imp}$

We already computed $\langle|H'_{kk'}|^2\rangle_{imp}$ in the main text. In the computation of $\langle(|H'_{kk'}|^2)^2\rangle_{imp}$ one will have four integrals

$$\begin{aligned}\langle(|H'_{kk'}|^2)^2\rangle_{imp} &= \frac{|\phi_{kk'}|^4}{V^4}\sum_{i,j,k,l}\left(\frac{1}{V}\int d^3r_i e^{-ir_i.q}\right)\left(\frac{1}{V}\int d^3r_j e^{ir_j.q}\right)\\ &\times \left(\frac{1}{V}\int d^3r_l e^{-ir_l.q}\right)\left(\frac{1}{V}\int d^3r_m e^{ir_m.q}\right)\end{aligned} \tag{8.5}$$

Respecting the condition that $q \neq 0$, one will have nonzero terms with $i = j$ and $l = m$ or with $j = l$ and $i = m$ or with $i = j = l = m$. Write these conditions in terms of delta functions as $\delta_{ij}\delta_{lm} + \delta jl\delta_{im} + \delta_{ml}\delta_{lj}\delta_{ji}$. On implementing these in the summations and performing the summations leads to

$$\langle(|H'_{kk'}|^2)^2\rangle_{imp} = \frac{|\phi_{kk'}|^4}{V^4}(2N^2 + N). \tag{8.6}$$

From the knowledge of $\langle|H'_{kk'}|^2\rangle_{imp}$ and the above, rms can be shown to be ~ 1 in the large N limit, i.e.,

$$\frac{\sqrt{\langle(|H'_{kk'}|^2)^2\rangle_{imp} - (\langle|H'_{kk'}|^2\rangle_{imp})^2}}{\langle|H'_{kk'}|^2\rangle_{imp}} \sim 1. \tag{8.7}$$

8.3 Appendix C: The concept of quasiparticles

In this appendix we will discuss the basic ideas of a very famous theory due to Lev Landau: the existence of quasiparticles and Fermi Liquid Theory (FLT).* Discussion is qualitative and emphasis is on the notion of quasiparticles. At the end of it we will learn how the spectral function captures the interaction effects.

From experience we learn that in metals electrons can be treated as free (up to a reasonably good approximation). This must be a very counterintuitive and non-trivial observation. The following consideration will show that in metals electrons should form a strongly interacting gas. Electrons interact with each other via Coulomb interactions. At metallic densities (say each atom liberates one electron) the mean separation between any two electrons is of the order of lattice spacing ($a \sim 1\mathring{A}$). At this distance electron-electron Coulomb potential energy is of the order of several electron volts. The kinetic energy, i.e., the Fermi energy, is also of the similar order (several electron volts). Thus in metals the kinetic and potential energies are of the same order (at this smaller distance the screening effects will not change the picture drastically) and electrons form a strongly interacting electron gas. Therefore, the free electron model is quite counterintuitive. How is it possible to treat an interacting electron gas as free (non-interacting) electron gas?

Thus we confront a daunting problem here. It was Lev Landau's powerful intuition† that recognized that *due to phase space limitations, the electron-electron scattering is suppressed, and momentum actually remains a good quantum number which justifies the notion of quasiparticles. Due to Pauli principle an electron can scatter to a state only if it is unoccupied. At suffi-*

* "You cannot repeal Coulomb's law," Lev Landau (page 86, quoted by David Pines in *BCS: 50 Years*, Ed. by D. Feldman and L. N. Cooper (2010)).

† A. A. Abrikosov and I. M. Khalatnikov, *Rep. Prog. Phys.* **22**, 329 (1959).

ciently low temperatures (compared to Fermi energy) this restriction becomes very severe and electron-electron scattered is suppressed. It turns out that the electronic mean free path becomes much bigger than lattice constant, by several orders of magnitude ($\Delta l >> a$), and the uncertainty in momentum $\Delta p \sim \frac{\hbar}{\Delta l} << p_F$ ($p_F \sim \frac{\hbar}{a}$).

Thus momentum remains a good quantum number and it becomes meaningful to define quasiparticles with well-defined momentum. Based on these phase space restrictions it can be argued (and can be proved with rigor) that there is a one-to-one correspondence between interacting electrons (via Coulomb interaction) and weakly interacting quasiparticles. This is the basis of adiabatic continuity and a system of strongly interacting electrons can be replaced with weakly interacting quasiparticles. This adiabatic continuity is not generally true (e.g., in fractional quantum Hall effect it is violated). But in 3D metals it is generally true and the weaker scattering mechanisms remain, and one can further define an effective lifetime to an occupied state. In the following we will incorporate these effects in the retarded Green's function (Green's function of the free particles case) and observe how the physical situation is changed by considering a modified spectral function.[‡] The treatment is phenomenological and serves a qualitative guide to actual physical calculations of Green's functions in realistic many-body systems.

For free particles the Green's function is given by the equation

$$G_R(\mathbf{k}, t) = -ie^{i\epsilon_{\mathbf{k}} t}\Theta(t). \tag{8.8}$$

The amplitude oscillates in time but never decays. If there are interaction effects then there is a mean lifetime to a given state. If we assign a phenomenological lifetime to state $\mathbf{k}$ by $\tau_{\mathbf{k}}$, then the modified Green's function takes the form

[‡]For introduction to Green's function refer, for example, to Rickayzen, *Green's Functions and Condensed Matter*, Dover Books, reprinted (2013).

$$G_R(\mathbf{k},t) \simeq -ia_{\mathbf{k}} e^{-i\tilde{\epsilon}_{\mathbf{k}} t} e^{-t/\tau_{\mathbf{k}}}. \tag{8.9}$$

Here two more interaction effects are taken into account. Interactions not only give a finite lifetime to a state but also they renormalize the energy as $\epsilon_{\mathbf{k}} \to \tilde{\epsilon}_{\mathbf{k}}$ and amplitude by a factor of $a_{\mathbf{k}}$. Due to finite lifetime, probability amplitude reduces to a number less than one. In rigorous theory we will notice that the renormalization factor, change in energy due to renormalization $\tilde{\epsilon} - \epsilon$, and lifetime are not independent from each other but rather connected with each other through general principles (like causality).

Fourier transform of Equation (8.9) takes the form

$$G_R(\mathbf{k},\omega) = \frac{a_{\mathbf{k}}}{\omega - \tilde{\epsilon}_{\mathbf{k}} + i\frac{1}{\tau_{\mathbf{k}}}}. \tag{8.10}$$

And the spectral function

$$A(\mathbf{k},\omega) = -\frac{1}{\pi} Im G_R(\mathbf{k},\omega) = \frac{1}{\pi} \frac{a_{\mathbf{k}}(1/\tau_{\mathbf{k}})}{(\omega - \tilde{\epsilon}_{\mathbf{k}})^2 + \frac{1}{\tau_{\mathbf{k}}^2}}, \tag{8.11}$$

has the form of a Lorentzian (a broadened delta function). An interesting point to note is that if the lifetime of state is infinitely large (no-scattering case) then the above Lorentzian spectral function reduces the delta function (as can be checked by taking the limit in the above expression). Thus, interactions make the spectral function broader. For example, consider that spectral functions are measured (by some experimental method[§]) in two systems: system A and system B. Suppose these measured spectral functions are shown schematically in Figure 8.3. For the figures we notice that spectral function for system A is broader than that in system B. From this observation we can conclude that system A contains more strongly interacting particles and the width at half of the maximum of the spectral function measures the inverse lifetime. Thus, quasi-particle lifetime will be shorter in system A.

[§]Generally, one can measure spectral function with photo-emission experiments.

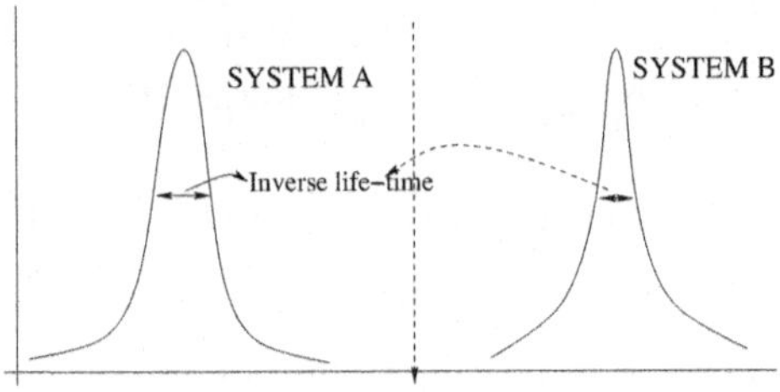

FIGURE 8.1: Schematic spectral functions in two systems A and B.

8.4 Practice exercises

1. Prove that the reactive component in Equation (1.47) does not lead to dissipation.

2. When the electric field is given by $E = E_0^x e^{\epsilon t}\Theta(-t)$, prove Equation (2.22), i.e, current density exponentially decays with time with a time scale which is of the order of relaxation time of the system.

3. Prove the formula for AC conductivity (Equation (2.24)).

4. Derive Equation (2.21) from Equation (2.34)[use $\mathbf{F} = -e\mathbf{E}$]. There is no explicit time dependence in distribution function and system is spatially homogeneous.

5. Show that the scattering rate for elastic scattering due to impurities in an isotropic medium

$$\frac{1}{\tau(\mathbf{k})} = \int \frac{d^3k'}{(2\pi)^3} W_{\mathbf{k},\mathbf{k}'}(1 - \cos(\theta)) \tag{8.12}$$

reduces to

$$\frac{1}{\tau(\mathbf{k})} = \frac{m}{\pi\hbar^2} n_i U^2 |\mathbf{k}| \tag{8.13}$$

for a simple metal with spherical Fermi surface and $W_{\mathbf{k},\mathbf{k}'} = \frac{2\pi}{\hbar} n_i \delta(\xi(\mathbf{k}) - \xi(\mathbf{k}'))|U_{\mathbf{k},\mathbf{k}'}|^2$ in the impurity scattering case.

6. Plot $f(\eta) = \frac{1}{e^{\eta}+1}$ and $h(\eta) = \frac{1}{e^{\eta+z}+1}$ for various values of z and notice how the area enclosed between these two curves goes with z, and then show that

$$\int_{-\infty}^{+\infty} d\eta \frac{1}{(e^{\eta+z}+1)(1+e^{-\eta})} = \frac{z}{e^{z}-1}. \tag{8.14}$$

7. Derive the Drude formula of conductivity from Equation (3.48) under the approximation of exponential decay of the current-current correlation function.

8. Follow the treatment given in Section 3.6 on generalized Langevin equation and show that

$$m\gamma[\omega] = \frac{1}{k_B T}\int_0^{\infty} dt \langle R(t_0)R(t_0+t)\rangle e^{-i\omega t}, \tag{8.15}$$

and then show that conductivity can also be expressed in terms of force-force correlation function. [Hint: Start with RHS. Substitute $R(t+t_0) = m\dot{u}(t+t_0) + m\int_{t_0}^{t_0+t} dt'\gamma(t-t')u(t')$]

9. Derive expressions for AC mobility

$$\mu[\omega] = \frac{1}{m}\frac{1}{i\omega + \gamma[\omega]} \tag{8.16}$$

for sample memory functions: (a) $\gamma(t) = \gamma_0 e^{-\alpha t}$ and (b) $\gamma(t) = \gamma_0^3 t e^{-\alpha t}$.

10. Show that the imaginary part of the memory function for lowest order spin-flip scattering of electrons over impurities with spin with interaction Hamiltonian $H_{electron-spin} = \frac{1}{N}\sum_J U^J_{\mathbf{k},\mathbf{k}'} S^i_J s^i_{\sigma\sigma'} c^{\dagger}_{\mathbf{k},\sigma} c_{\mathbf{k}',\sigma'}$ is given by

$$M''(\omega) = C[S(S+1) + \frac{1}{2}\langle S_0\rangle \left(\frac{\omega-h}{\omega}\coth\frac{\omega-h}{2T} - (h\to -h)\right)]. \tag{8.17}$$

Refer to W. Götze and P. Wölfle, *Phys. Rev. B.* **6**, 1226 (1972).

11. Evaluate

$$\phi_{\mathbf{k}\mathbf{k}'} = \frac{1}{V}\int d^3r e^{-i(\mathbf{k}-\mathbf{k}')\cdot\mathbf{r}}\phi(\mathbf{r}), \tag{8.18}$$

when $\phi(\mathbf{r}) = \phi_0 e^{-r/\xi}\cos(\alpha r)$ where $\xi << L$. Plot $\phi_{\mathbf{k}\mathbf{k}'}$ or $H'_{\mathbf{k}\mathbf{k}'}$ as a function of $q = |\mathbf{k}-\mathbf{k}'|$ for various choices of ξ and α.

12. In Figure 8.2 a Mathematica program is given. The purpose of it is to show that the sum ($S = \sum_{i=1}^{N} e^{-i(k-k').r_i}$) goes as ($\sqrt{N}$) when the positions of the impurities take on random values.

```
(*Off-Diagonal Elements are suppressed
 due to random impurities*)
lc = 1; (*lattice constant =1*)
xn = 10 ^ (6); (*total number of impurity sites.*)
xls = 10 ^ (14); (*total number of lattice sites*)
tn = 39 000; (*an integer for wavevector*)
xk = 2 π tn; (*wave vector k, tn is an integer*)
su = 0;
Do[
  y = RandomInteger[{0, xls}, {xn}];
       xn
  u =  Σ  Exp[-I xk 1.0 (y[[i]])];
      i=1
  su = su + u;
  , {j, 1, 100, 1}];
ssu = su / 100 (*Calculted Average*)
s = N[√xn] (*√N , exact average*)


979.647 - 5.9635 i

1000.
```

FIGURE 8.2: A Mathematica program that shows that the off-diagonal goes as $\sqrt{N}$.

Study the program, and then execute it by changing (a) the number of impurity atoms, (b) total number of lattice sites, and (c) for different wave-vector $k - k'$ by changing the integer tn.

13. Define

$$M = \frac{1}{\nu} \sum_{k' \in \nu} |H'_{kk'}|^2, \tag{8.19}$$

where the sum over k' (in Equation (5.33)) is divided into patches of size ν and ν is a macroscopic number of the order of N. Show that

$$\langle M \rangle_{imp} = \frac{|\phi_{kk'}|^2}{V^2} N$$
$$\langle M^2 \rangle_{imp} = \frac{|\phi_{kk'}|^4}{V^4} [N^2 + \frac{1}{\nu^2} N(N-1) \sum_{k',k'' \in \nu} (\delta_{k'k''} + \delta_{2k,k'+k''})] \quad (8.20)$$

and the rms fluctuations goes to zero in the large N limit [Hint: refer to Appendix 8.2].

14. Show that the conductivity $\sigma = \frac{ne^2\tau}{m}$ can be written as

$$\sigma_{2D} = \frac{e^2}{h}(k_F l), \quad (8.21)$$

when $n = 2\int_0^{k_F} \frac{d^2k}{(2\pi)^2}$ in 2D. Find out the value of the dimensionless parameter $k_F l$ for the compound $(La - Sr)CuO_4$ if $\rho = 0.05 m\Omega\ cm$ at $T = 50\ K$. Inter-plane distance is 6 $\mathring{A}$.

Index

Absorption coefficient, 18
Adiabatic continuity, 204
Adiabatic switching on, 47
Admittance, 80, 85
ADS/CFT correspondence, 197
Anomalous skin effect, 191
Average time for acceleration, 4

Back scattering, 54
Bad metals, 160
Beyond the relaxation time approximation, 49
Bloch waves, 39
Bloch's assumption, 56
Bloch-Boltzmann transport equation, 54
Boltzmann equation, beyond RTA, 51
Bottleneck, in momentum transfer, 181
Brownian motion, 96, 99

Callen-Welton theorem, 76
Central limit theorem, 96
Chamber's method, 42
Collision integral, 50, 51
Colloidal suspension, 94
Colors of metals, 30
Cuprates, 152

DC conductivity, 4
DC resistivity, strange metals, 155
Debye approximation, 129
Density matrix, 83
Dielectric function, 17
Diffusion coefficient, 94
Distribution function, 41
Drude model, 16
Drude peak, 168
Drude's formula, 46, 88

Einstein relation, 94, 95
Electron wavepackets, 40
Electron-lattice interaction, 54
Ellipsometry, 16
Ergodicity, 44

Fermi's Golden rule, 55, 57
Fluctuating magnetic substratum, 158

Fluctuation-dissipation theorem, 75, 82, 88
Forward scattering, 54
Fourier's law, 6

Götze-Wölfle (GW) formalism, 116
Generalized Drude formula, 104
Generalized Drude scattering rate, 166
Generalized Langevin equation, 102
Grüneisen formula, 68, 130
Gros-Rice transformation, 178
Gutzwiller projection, 173
Gutzwiller projector, 178

Hagen-Rubens regime, 23
Hidden Fermi liquid theory, 172
Holstein mechanism, 125
Holstein-Boltzmann equation, 193
Hot electrons, 192
Hot spots and cold spots heuristics, 172

Impurity scattering, 120, 139
Insulators, 25

Johnson-Nyquist noise, 73

Kohn-Luttinger theory, 135
Kramers-Kronig relations, 92
Kubo's formula for conductivity, 86

Landau Fermi liquid theory, 190, 203
Langevin equation, 98
Linear response theory, 82
Local thermodynamic equilibrium, 51
Logarithmic divergence in resistivity, 162
Lorentz number, 7
Lorentz oscillator model, 25

Marginal Fermi liquid theory, 172, 174
Markovian approximation, 87
Mean collision time, 46
Memory function, 104
Memory matrix, 115
Mesoscopic systems, 192
Microscopic reversibility, 51
Mid-infrared band, 167
Mobility, 97
Mott-Ioffe-Regel limit, 159

Nearly anti-ferromagnetic Fermi liquid theory, 172
Non-Fermi liquid theories, 173
Nonequilibrium distribution function, 44
Nyquist's theorem, 73

Ohm's law, 2
Optical conductivity, cuprates, 167
Optical properties of metals and insulators, 29

Phase diagram, cuprates, 153
Phase space, 42
Phonon drag, 191
Phonon scattering, 125
Photoemission, 205
Planckian dissipation limit, 185
Projection operators, 112
Pseudogapped regime, 154, 160
Pseudoparticles, 173, 174, 179

Quantum criticality, 172, 173, 184
Quasiparticle weight, 173, 175

Random phase assumption, 138, 191
Reduced density matrix, 141
Reflectivity and absorptivity, 19
Refractive index, 20
Relaxation function, 85
Relaxation time, 48
Relaxation time approximation, 42
Response function, 84

Semiclassical model, 38
Skin effect, 23
Sommerfeld's contribution, 9
Spectral function, 205
Spin-flip scattering, 207
Strange metals, 152
Survival probability, 44

t-J model, 178
T-linear resistivity, 155
Thermal conductivity, 5
Tube-integrals, 43

Unconventional superconductors, 194
UV transparency, 28

Velocity-velocity correlation function, 97

Weber's model, 2
Wiedemann-Franz law, 3, 7

X-ray edge singularity problem, 180

01- 6000- 1